M. I. Budyko   A. B. Ronov   A. L. Yanshin

History of the Earth's Atmosphere

With 40 Figures

Springer-Verlag
Berlin Heidelberg New York
London Paris Tokyo

Professor Dr. Michael I. Budyko
Professor Dr. Alexander B. Ronov
Professor Dr. Alexander L. Yanshin
State Hydrological Institute
23. Second Line, Basil Island
Leningrad 199053, USSR

Translated from Russian by:
S. F. Lemeshko and V. G. Yanuta

Title of the original Russian edition:
Istoriya Atmosferi
© by Gidrometeoizdat, 1985

ISBN 3-540-17235-1 Springer-Verlag Berlin Heidelberg New York
ISBN 0-387-17235-1 Springer-Verlag New York Berlin Heidelberg

© Springer-Verlag Berlin Heidelberg 1987
Printed in Germany

Typesetting: Overseas Typographers, Inc., Makati, Philippines
Offsetprinting and Bookbinding: Druckhaus Beltz, Hemsbach/Bergstraße
2132/3130-543210

Preface

The authors of this book have studied the changes in the chemical composition of the atmosphere during geological history with regard to its close relationship to the evolution of the Earth's sedimentary shell.

Beginning in 1977, the initial results of this study have been published as articles and parts of several monographs. Since new material clarifying atmospheric evolution have been obtained recently, the necessity has arisen to write a book treating the major results of investigations of the history of the atmosphere. In this book much consideration is given to the interrelation between the evolution of animate nature and changes in atmospheric composition. It proved be necessary to study the history of the two components of atmospheric air: carbon dioxide and oxygen. Attempts have been made to represent quantitatively the conclusions drawn here, i.e. to determine by calculation the changes in the amount of carbon dioxide and oxygen over much of the history of the atmosphere.

These calculations, performed in most detail for the Phanerozoic and to a lesser degree for the Late Precambrian, are supplemented with estimates of changes in the chemical composition of the atmosphere in the Early Precambrian.

Comparisons have been drawn between the changes in the chemical composition of the atmosphere and the development of animate nature, a close relationship being found to exist between the stages of the evolution of organisms and variations in the chemical composition of the atmosphere.

The data on the variations in amount of carbon dioxide during geological history can be used to study the contemporary anthropogenic climate change caused by the growth of atmospheric CO_2 due to the burning of increasing amounts of coal, oil and other kinds of carbon fuel. The climatic conditions of those epochs when the atmosphere contained great amounts of carbon dioxide can be considered as analogues of the future climate. This is frequently used when estimating the climatic changes expected to occur by the end of the 20th and in the 21st century.

The sensitivity of the biosphere to external factors is treated in the closing section of the book. The conclusion is drawn that the maintenance of comparative stability in the Earth's climate, or any change of atmospheric chemical composition within whose limits the biosphere can still survive are the consequence of the hardly probable coincidence of several factors, independent of each other, in the biosphere's evolution. It is thus assumed that the Earth's atmosphere (as well as the biosphere) is a unique phenomenon in that part of the universe closest to the Earth.

M. I. BUDYKO
A. B. RONOV
A. L. YANSHIN

Contents

1   Introduction   1

1.1 The Modern Atmosphere   1
1.2 Cycles of Atmospheric Gases 11
1.3 Studies of the Evolution of the Atmosphere 17

**2   Methods for Determining Changes in the Composition
of the Atmosphere** 33

2.1 Sedimentary Layer of the Earth's Crust 33
2.2 Carbon in the Sedimentary Layer 52
2.3 The Dependence of Amounts of CO_2 and O_2 in the
Atmosphere on Carbon Mass in Sediments 64

**3   The Evolution of the Chemical Composition of the
Atmosphere** 79

3.1 Carbon Dioxide 79
3.2 Oxygen 98
3.3 Past and Future of the Atmosphere 118

Conclusion 127

References 131

Subject Index 137

1 Introduction

1.1 The Modern Atmosphere

Chemical Composition of the Atmosphere. The atmosphere consists of a mixture of gases. Dry air contains mainly nitrogen, oxygen, argon and carbon dioxide. The amounts of these gases in dry air and their molecular masses are presented in Table 1.

Table 1. Chemical composition of the atmosphere

Gas	Volume concentration (%)	Molecular mass
Nitrogen	78.08	28.0
Oxygen	20.95	32.0
Argon	0.93	39.9
Carbon dioxide	0.03	44.0

Besides these, dry atmospheric air contains small quantities of neon, helium, methane, krypton, hydrogen and some other gases. The mean molecular mass of dry air is equal to 28.96. The relative concentrations of atmospheric air components included in Table 1 change comparatively little in different geographical regions and do not in practice depend on the altitude in the lower atmospheric layers (up to several tens of kilometres). This does not apply to the remaining large gas component of the atmosphere, water vapour, the amount of which in the surface layer is within 0.1-1%, sometimes decreasing to hundredths of a per cent or increasing to several per cent. Water vapour concentration decreases with height, being insignificant as a rule above the 10 to 20 km level.

We will consider additionally only those atmospheric components that produce pronounced effects on physical and chemical processes in the atmosphere and on living organisms. Among these are all the gases with concentrations exceeding 0.01% except argon, which belongs to the group of inert gases and scarcely influences the processes in the atmosphere and the biosphere. Thus, the major components of atmospheric air include nitrogen, oxygen, carbon dioxide and water vapour.

The total mass of the atmosphere is approximately 5.2×10^{21} g, equivalent to about 10^3 g cm^{-2} for the unit area of the Earth's surface. Masses of the major gas components in dry atmospheric air equal: 3.9×10^{21} g for nitrogen, 1.2×10^{21} g for oxygen, and 2.6×10^{18} g for carbon dioxide. The mass of water vapour makes up 0.12×10^{21} g, which corresponds to 2.4 g cm^{-2}.

Three of the above gases, nitrogen, carbon dioxide and water vapour, enter into the atmosphere from the depths of the Earth during the degassing of the Earth's mantle and crust. These gases are amongst those that are ejected into the atmosphere by volcanoes from the deep fissures in the Earth's crust, and from hot springs. The amount of oxygen in volcanogenic gases is insignificant. The photosynthesis of green plants is the main process providing oxygen into the atmosphere. During photosynthesis, carbon dioxide and water form carbohydrates producing free oxygen.

All the major gas components of the atmosphere are consumed and re-enter the atmosphere after interacting with organisms, products of their life activity, water of the hydrosphere and some mineral substances. The distinctive features of these cycles are covered in the second section of this chapter.

Here we present brief information on each of these four components.

Nitrogen is one of the chemical elements extensively distributed in the Solar System. It is less chemically active than other major components of air, so that the rate of its circulation is relatively slow. Atmospheric nitrogen basically influences climate by its greater mass (compared to other atmospheric gases), determining to a considerable extent air density and pressure at different altitudes. Therefore, nitrogen has indirect but important effects on general atmospheric circulation.

As a result of electric discharges in the atmosphere, nitrogen molecules oxidize and form NO, which on cooling yields NO_2. Nitrogen oxides are also produced in photochemical reactions in high atmospheric layers and by the activity of nitrifying bacteria. (See next section). Although the total mass of these oxides is insignificant, they affect some atmospheric processes mainly through their effects on the ozone in the atmosphere. An increase in concentration of nitrogen oxides can lead to a decrease in ozone mass and an attenuation of the ozone screening effect, which protects living organisms against ultraviolet radiation.

Nitrogen is important for living organisms because it enters into proteins and various proteinaceous compounds. The overwhelming majority of living organisms cannot assimilate nitrogen from the atmosphere, where its content is practically unlimited, but consume nitrogen produced mainly by nitrogen-fixing bacteria and algae.

Oxygen, like nitrogen, is extensively distributed in the Solar System. Unlike nitrogen, it is chemically very active. Oxygen and its compounds comprise a major part of the external envelope of the Earth: the hydrosphere, much of the lithosphere and a noticeable portion of the atmosphere. Unlike the hy-

drosphere and the lithosphere, the atmosphere contains for the most part un-
bonded oxygen.

It is paradoxical that there is a great quantity of an element as active as oxygen
in the modern atmosphere; this can be attributed to the high rate of its entry into
the atmosphere as a result of its accumulation in sediments of organic carbon
formed by green plants.

In addition to diatomic oxygen (O_2), the atmosphere contains three-atom
oxygen (O_3), ozone, which is formed mainly at heights of 10 to 50 km as a result
of the effects of ultraviolet solar radiation on the oxygen molecules. Ozone has a
physical effect on the atmosphere (e.g. the air temperature increases in the layers
containing the greatest amounts of ozone). It is also very important for organisms,
as it detains a great proportion of the ultraviolet radiation dangerous for many
living beings.

Atmospheric oxygen is transferred by turbulent and molecular diffusion into
water bodies where it dissolves. The quantity of oxygen in water bodies depends,
among other factors, on their temperature. When photosynthesizing plants are
present in the upper layers of water bodies, oxygen produced in photosynthesis
enters them additionally. The total amount of free oxygen in reservoirs is far less
than its mass in the atmosphere.

Oxygen is of vital importance for living organisms. The overwhelming majority of organisms obtain energy for their life from oxidizing organic matter. Dead organic matter is also destroyed by oxidation. Thus, great
amounts of oxygen are consumed in oxidizing organic matter, rocks and
reducing gases (carbon monoxide etc.) entering the atmosphere from the depths
of the Earth.

There is a relatively small amount of carbon dioxide in the modern atmosphere; however, this gas has a pronounced effect on climate. Since carbon dioxide
is practically transparent for shortwave solar radiation and absorbs a considerable portion of longwave heat radiation, it is important in producing the
greenhouse effect that increases the air temperature at the Earth's surface. In this
sense, the importance of carbon dioxide is comparable with that of water vapour.

Without carbon dioxide life on Earth would be impossible. It is essential for
organisms because all living matter produced in the modern biosphere is formed
almost entirely from carbon dioxide and water during photosynthesis. Since the
productivity of photosynthesis usually depends on the concentration of CO_2,
changes in this concentration can lead to variations in the total mass of organisms
on the Earth.

Water bodies contain quantities of CO_2 much greater than the atmosphere,
its total mass reaching 130×10^{18}g. The carbon dioxide of the hydrosphere is used
by water plants in photosynthesis. Organic matter formed in green plants is
consumed by heterotrophic organisms which could not have existed without this
source of energy.

The carbon dioxide of the atmosphere has a pronounced effect on the erosion
of rocks. When dissolving in land surface waters, CO_2 becomes a chemically

active agent accelerating the process of chemical weathering of silicate minerals, in the course of which calcium and magnesium compounds that form carbonate sediments in water bodies are produced.

Water vapour, the fourth major component of air, differs from the three others in some respects. One distinction is that nitrogen, oxygen and carbon dioxide exist only as gases under past and present climatic conditions. By contrast, gaseous water (water vapour) in the modern atmosphere comprises the smallest part in mass of the free water existing in the hydrosphere and the cryosphere mainly in a liquid state (the mass of liquid water is 1.4×10^{24} g) and in a solid state (the mass of ice and snow is 3×10^{22} g). At the same time, while the amounts of nitrogen, oxygen and carbon dioxide in the atmosphere can vary within wide limits, the volume of water vapour is basically determined by solar radiation influx and air temperature and cannot vary considerably when these factors are constant. This can be explained by the fact that water bodies occupy more than 70% of the Earth's surface, and that the greater part of the continents are kept humid by precipitation. Thus, an appreciable amount of water evaporates annually from the Earth's surface into the atmosphere, the rate of evaporation depending mainly on the solar radiation regime and air temperature.

As a result, the atmosphere receives much water vapour which is transferred by air fluxes to higher tropospheric layers, where it condenses and then precipitates. The relative amount of water vapour in the atmosphere (relative air humidity) is comparatively permanent if its cycle remains more or less constant. Under these conditions, the amount of water vapour (absolute air humidity) depends only on temperature.

Climate has varied noticeably in the geological past, therefore the amount of water vapour in the atmosphere has also changed. However, variations in water vapour content were the consequence and not the cause of climate change.

This conclusion does not contradict the fact that water vapour has a pronounced effect on climate. Water vapour is minimally transparent for longwave heat radiation, therefore it is important in creating the greenhouse effect in the atmosphere. Climatic conditions are also influenced by the transference of water vapour by horizontal air flows, since in condensing water vapour and forming clouds an appreciable amount of heat is released, which increases the air temperature in corresponding geographical regions.

Water vapour definitely influences the vital activity of organisms. For instance, the transpiration rate (the loss of water by plants in evaporation) depends on air humidity, increasing with decreasing humidity. However, the direct influence of water vapour on organisms is less essential than its indirect effects as the source of water falling on the continents as precipitation.

A part of the total volume of water vapour produced by evaporation from the surfaces of oceans and other water bodies is transferred by horizontal air flows to the continents, where precipitation forms and falls as rain, hail and snow. Precipitation on the continents humidifies soil, which makes the existence of plants possible. Thus, without water vapour in the atmosphere, no living organism could exist on the continents.

Moreover, water vapour enters the atmosphere from water bodies, so that an atmosphere without water vapour can exist only on a planet deprived of liquid water, and thus unsuitable for organic life.

Comparing the effects of the major components of atmospheric air (nitrogen, oxygen, carbon dioxide and water vapour) on climate and living nature, we can conclude that these effects are not in proportion to their relative volumes.

Since organic matter consists of carbon compounds, it is evident that without the carbon dioxide cycle life on Earth would be impossible. Taking this into account and considering that variations in CO_2 concentration in the atmosphere have appeared to be the main cause of climatic change in the geological past, it is easy to understand that carbon dioxide, a small component of the atmospheric air, is of primary importance in the history of the atmosphere and the biosphere.

As mentioned above, free oxygen is the most important factor, providing the life activity of modern organisms.

However, in the remote past life developed over a long period in an atmosphere deprived of oxygen. Free oxygen is therefore not a condition for the existence of the biosphere.

Variations in the amount of atmospheric nitrogen were less important for animate nature because the nitrogen concentration throughout the history of the atmosphere has much exceeded the amount that could have been utilized by organisms. At the same time, the climatic effects of atmospheric nitrogen were comparatively limited.

While these three gases could act as external factors influencing atmospheric processes, water vapour, when the hydrosphere was present, was a component of a climatic system determined entirely by conditions which were external, relative to climate. Thus, variations in water vapour concentration were not an independent factor that affected the history of the atmosphere and the biosphere.

This book therefore, covers basically the history of variations in the amounts of carbon dioxide and oxygen in atmospheric air. Variations in atmospheric nitrogen mass are discussed briefly.

Climate. The atmosphere is characterized not only by its chemical composition but also by its physical state, i.e. the major meteorological elements determining the climatic conditions over various regions of the globe.

Air density and pressure decrease with height, the pressure decreasing by half with an increase in height of about 5 km. The lower air layer, the troposphere, extends to the height of 8 to 10 km in polar regions and to 16-18 km in tropics. In the troposphere, air temperature usually drops, at an average rate as height increases by $6°C\,km^{-1}$. Above the troposphere is the stratosphere, where up to the level of approximately 50 km, air temperature increases with height. The troposphere contains more than two thirds of the atmospheric volume.

Solar radiation is almost the only source of energy for all physical processes occurring in the atmosphere. The greenhouse effect is very important in the radiation of the atmosphere which absorbs little shortwave solar radiation, this mainly reaching the Earth's surface, but retains longwave heat radiation, thus increasing the temperature of the lower air layers.

The causes of the differences in climate in different regions of the Earth already attracted the attention of philosophers in antiquity. Even at that time it was known that climatic conditions were closely connected with the average altitude of the Sun, i.e. the latitude of different areas.

Modern theories of the physical mechanism of the Earth's climates can be interpreted as follows. Differences in radiation regime due to the spherical form of the Earth result in an uneven heating of the atmosphere over different regions of the globe, differences in surface temperature between the equator and the poles being particularly great. This is the major cause of atmospheric circulation. Topography, including the location of the continents and the oceans, also greatly influences the motion of atmospheric air.

Due to their high heat capacity and conductivity, oceanic waters have great thermal inertia, i.e. they can considerably reduce temperature variations occurring, for example, as a result of annual changes in solar radiation production. Due to this, the air temperature over the oceans in middle and high latitudes is much lower in summer and higher in winter than over the continents. Lack of uniformity in heating the atmosphere leads to the emergence of the complicated system of air flows called general atmospheric circulation. This provides for horizontal and vertical heat transfer in the atmosphere, as a result of which differences in heating the atmosphere over individual regions are noticeably levelled out. At the same time, general circulation maintains the water exchange in the atmosphere: water vapour is transferred from the oceans to land.

The motion of atmospheric air is determined by the distribution of pressure, which itself depends on atmospheric circulation. At sea level, pressure decreases at the equator and in middle latitudes and increases in the subtropics (band of high pressure). At the same time pressure over the continents in extratropical latitudes usually increases in winter and is lower in summer.

The pressure distribution greatly influences the system of air flows, some of which are comparatively time-independent, others varying permanently with space and time.

The trade-wind system is one of stable air flows. Trade-winds blow from subtropical latitudes to the equator in both hemispheres, and can be distinctly observed over the oceans. Monsoons, the seasonal air flows between the oceans and the continents, are also comparatively stable. Monsoon circulation occurs in the eastern and southern periphery of Asia, in equatorial Africa and in some other regions.

In middle latitudes western air flows prevail. They are transient and include large eddies, with the cyclones and anticyclones usually covering hundreds of kilometers.

The tropical cyclones cover a smaller area, but have particularly high wind velocities, comparable with those of storms.

In the upper troposphere and the lower stratosphere, peculiar jet streams occur with sharp boundaries within which the wind velocity ranges from 150 to 200 km h^{-1}.

The climatic conditions in different regions depend greatly on the water cycle in the atmosphere. The water cycle includes evaporation from the oceans and the continents, transfer of water vapour in the atmosphere, its condensation, precipitation and river run-off.

While total global precipitation is equal to evaporation from the oceans and continents, in individual regions evaporation and precipitation differ greatly in quantity. In this case evaporation from land is usually less than total precipitation.

Because of this a part of the precipitation over land forms from water vapour transferred by air flows from the oceans. The difference between evaporation and precipitation on the continents is equal to the difference between gain and loss in atmospheric water vapour over the continents, at the same time being equal to river run-off from the continents into the oceans.

The values of atmospheric water balance components are given in Table 2 (The World Water Balance. . . 1974).

The difference between land and ocean in gain and loss of water vapour in this table is explained by the fact that they refer to a unit area (for the oceans and land, as a whole, these values are the same).

Table 2. Water balance of the atmosphere (cm yr^{-1})

	Precipitation	Evaporation	Gain or loss of water vapour
Earth's atmosphere	113	113	
Over land	80	49	31
Over oceans	127	140	-13

Although the amount of water vapour coming from the oceans to the continents comprises a smaller portion of continental precipitation, this precipitation is in fact formed largely from water vapour transferred from the oceans. This is because a considerable part of the water vapour formed due to local evaporation is transferred by atmospheric circulation that brings to the continents much more water vapour from the oceans than is given in Table 2. The value in Table 2 represents the difference between gain and loss of water vapour.

The variety in the Earth's climate can be attributed to different amounts of solar radiation at different latitudes and to the structure of the Earth's surface, including the distribution of the oceans and the continents, and its topography.

The tropical climate is characterized by high air temperatures (mainly from 25° to 30°C) changing only slightly throughout the year. In the equatorial band, heavy precipitation is usually recorded. This creates local conditions of excessive humidity. In tropical latitudes outside the equatorial zone, precipitation decreases and in the high pressure zone it becomes very slight. In this region there are many vast continental deserts.

In subtropical and middle latitudes, temperature changes over the year considerably, the difference between winter and summer temperatures being particularly great within the land masses far from the oceans. Thus, for example, in some regions of Siberia the temperature difference between the coldest and the warmest months is 70°C. Humidity conditions in these latitudes are very diverse and depend largely on atmospheric circulation.

In polar regions, where seasonal temperature changes are noticeable, temperature is low throughout the year, which promotes the advancement of ice cover over land and ocean.

Climatic conditions on our planet vary constantly. Information about these variations in the past can be obtained from three major sources. Instrumental meteorological observations give accurate data on climate, but cover a period of not more than 100 years, since instrumental observation began only in the second half of the last century. Some information on the climate over the period of thousands of years can be obtained from noninstrumental observations described in various historical sources.

Data on the climate of more remote epochs of up to hundreds of millions or billions of years ago can be obtained in palaeogeographical investigations, which, to determine past climatic conditions, use the relationships between meteorological conditions and the life of plants and animals, with information on hydrological and lithogenetic processes. Oxygen isotope evidence in remnants of living organisms helps to determine the temperature of their environment and is of great importance in studying palaeoclimates.

Table 3 presents a geochronological scale now widely accepted and used in this book in discussing past climates.

Studies of past climatic conditions are summarized as follows.

Precambrian climatic conditions are known only in outline. The existence of liquid water throughout a greater part of Precambrian time and tracks of several vast ice sheets show that the Earth's surface temperature was lower than water boiling-point, sometimes below zero over a part of the globe. Numerous primitive organisms existed during a greater part of the Precambrian. This allows us to assume that the Earth's surface temperature lay within a narrow range of 0° to 50°C, or probably even 10° to 40°C, favourable for primitive organic life.

In the first half of the Phanerozoic the climate was warmer than at present, although during that time glaciations sometimes occurred over a part of the land. By the mid-Phanerozoic (the end of the Carboniferous and the beginning of the Permian periods) considerable glaciation had advanced into the continents of the Southern Hemisphere. When the ice that existed at the beginning of the Permian period disappeared, the climate grew warmer over a long period of time and no new glaciation occurred until the mid-Cenozoic Era. During the Mesozoic and the first half of the Cenozoic, the temperature difference between low and high latitudes was comparatively small. At the same time, the tropical temperature was close to the present, and the mid- and high latitude temperature was much higher than that observed now. The development of the present considerable tempe-

Table 3. Geologic time scale

Eon	Era	Period			Epoch	Years before present (m.y.)
Phane-rozoic	Cenozoic	Quaternary				0-2
		Tertiary	Neogene		Pliocene	2-9
					Miocene	9-25
			Palaeogene		Oligocene	25-37
					Eocene	37-58
					Palaeocene	58-67
	Mesozoic	Cretaceous				67-133
		Jurassic				133-186
		Triassic				186-236
	Palaeozoic	Permian				236-282
		Carboniferous				282-346
		Devonian				346-402
		Silurian				402-435
		Ordovician				435-490
		Cambrian				490-570
Protero-zoic	Vendian					570-670
	Late Riphean					670-1100
	Middle Riphean					1100-1350
	Early Riphean					1350-1600
	Early Proterozoic					1600-2600
Archaean						2600-3500
Catarchaean						3500-4500

Note: The Riphean and the Vendian form the Late Proterozoic; the Late Proterozoic and the Phanerozoic form the Neogäikum.

rature difference between equator and pole began in the late Mesozoic Era. It was much less than at present until the beginning of the Quaternary period (in the Northern Hemisphere) and the mid-Cenozoic (in the Southern Hemisphere).

The glaciation that occurred in the Quaternary period in the Northern Hemisphere varied considerably. It increased several times, reaching middle latitudes and then again retreating to high latitudes. The last (Würm) glaciation in Eurasia terminated about 10,000 years ago. After this, permanent ice cover was retained in the Northern Hemisphere mainly in the Arctic Ocean and on the islands at high latitudes. More extensive glaciation exists in the Southern Hemisphere in the Antarctic and adjoining oceanic areas.

The last 100 years, recorded in instrumental observational data, are characterized by comparatively small climatic changes. The warming that took place in the first half of the twentieth century was especially noticeable in the 1920-30's. The modern climatic changes are more considerable in middle and particularly in high latitudes of the Northern Hemisphere.

From the theoretical models used to study the physical mechanism of climatic changes, the following conclusions have been drawn. It is easy to estimate a surface temperature increase due to the greenhouse effect (i.e. less atmospheric transparency for longwave radiation than for shortwave solar radiation).

In the absence of the atmosphere, the mean surface temperature is determined by the radiation equilibrium equation:

$$\delta \sigma T^4 = \frac{1}{4} S_0 (1 - \alpha). \tag{1}$$

The left-hand side of this equation is the value of longwave heat radiation and the right-hand side corresponds to absorbed shortwave radiation; δ is the coefficient characterizing the difference between the properties of the emitting surface and black body; σ the Stefan constant, T the surface temperature, S_0 the solar constant, i.e. the solar radiation flux through a unit area perpendicular to solar rays at an average distance between the Sun and the Earth; α the average albedo (reflectivity) of the Earth. The value of δ is approximately 0.98, α is about 0.30 and solar constant S_0 equals 1368 W m^{-2}.

Equation (1) shows that the mean temperature of the Earth without the atmosphere would be 255 K or -18°C. Since the mean surface air temperature from observational data is 15°C, it is clear that the greenhouse effect increases the air temperature by 33°C. As has already been noted, the greenhouse effect of the atmosphere is largely attributed to the absorption of longwave radiation by water vapour and carbon dioxide.

The conclusion that the climate is ambiguous is important in studying the genesis of the Earth's climate. In other words, taking the above-mentioned value of solar constant and present composition of dry air, another climate, the so-called "white Earth" climate could take place (Budyko 1971).

In studying the relationship between longwave radiation produced into space, surface air temperature and cloudiness, it has been found that the value of outgoing emission is $a + bT - (a_1 + b_1 T) n$ (W m^{-2}), where T is the surface air temperature in degrees Celsius, n the cloudiness in fractions of a unit, $a = 223$, $b = 2.2$, $a_1 = 47.8$ and $b_1 = 1.6$ are the dimensional coefficients.

Taking this relationship into account, we find from the energy conservation law that for the Earth as a whole

$$\frac{1}{4} S_0(1 - \alpha) = a + bT - (a_1 + b_1 T) n. \tag{2}$$

Equation (2) demonstrates that with modern albedo and cloudiness (the latter being 0.5 for the entire Earth), the mean air temperature is approximately 15°C, i.e. the value obtained from the observational data. In the case of continuous ice-snow coverage on the Earth's surface, its albedo considerably increases, due to the high reflectivity of this surface, and is then close to the Antarctic albedo of 0.6 to 0.7 according to observational data.

As can be seen from Eq. (2), in this case mean surface air temperature with the indicated albedo is equal to a very low value of -50° to -70°C. It is clear that at these low temperatures complete glaciation of the Earth could be permanent, i.e. with the modern solar constant and the chemical composition of the atmosphere, the "white Earth" climate could exist for an unlimited period.

How would the factors determining the Earth's climate have be modified to alter the modern climate to cause complete glaciation of the Earth? This question is of considerable importance. Climate model calculations showed that this would be possible by decreasing the solar constant by 5%. It is also probable that the "white Earth" would occur at the present solar constant if the mass of atmospheric carbon dioxide decreased to a very low level.

Equation (2) shows how climate changes with changing heat influx to the Earth's surface. In particular, by increasing or decreasing the solar constant by 1%, the mean air temperature increases (or decreases) by 1.5°C. Thus, the "white Earth" can arise at an initial mean surface temperature decrease of 7°-8°C.

Theoretical climate studies show that the dependence of surface air temperature on the atmospheric carbon dioxide content is logarithmic, the temperature changing by 3°C with an increase of 100% (or decrease of 50%) in CO_2 content. Taking this into account, we find that a 7.5°C decrease in mean temperature is equivalent to reduction in CO_2 concentration from 0.03% to a very small value. As can be seen from the above-mentioned calculation of the "white Earth" temperature variations, global glaciation emerges due to decreasing CO_2 content. This glaciation would be retained even if the modern concentration of carbon dioxide in the atmosphere were restored.

The mean temperature-solar constant dependence mentioned above shows that an increase in S_0 by several dozen per cent could result in raising the temperature to water-boiling point. Since during this warming the Earth's albedo would decrease due to the melting of the present ice-snow cover and the amount of CO_2 in the atmosphere would increase due to a reduction in oceanic CO_2 content, it is probable that such a rise in temperature would take place with a considerably smaller increase in solar constant.

Undoubtedly, a similar climate change would occur even with the modern solar constant if the rate of CO_2 production to the atmosphere from the depths of the Earth were great enough. Similar conditions apparently formed on Venus, where in spite of closer proximity to the Sun than Earth, the solar energy absorbed by Venus due to greater albedo is about the same as that absorbed by the Earth. However, since the atmosphere of Venus is very rich in carbon dioxide, its surface temperature is close to 700 K.

Thus the Earth's climate is characterized by a high sensitivity to changes in external factors.

1.2 Cycles of Atmospheric Gases

Carbon Dioxide. The atmosphere receives carbon dioxide from the depths of the Earth. It serves for the formation of different kinds of carbon deposits. The cycle of carbon dioxide involves CO_2 exchange between several reservoirs of carbon where the masses of carbon are very different.

The sedimentary shell of the Earth contains the greatest amount of carbon, about 100×10^{21} g (see Sect. 2.2). Nearly 130×10^{18} g of carbon dioxide, equivalent to 35×10^{18} g of carbon, are dissolved in the oceans and other water bodies.

Forest biomass contains a great part of the carbon entering living organisms, the total quantity of carbon in organisms being about 0.8×10^{18} g. The mass of organic carbon in the soils, also known only approximately, amounts to about 2×10^{18} g. The atmospheric carbon dioxide, the mass of which is 2.6×10^{18} g, contains about 0.7×10^{18} g of carbon.

The CO_2 exchange between these resources takes place in various forms which correspond to different rates of the CO_2 cycle. Some kinds of CO_2 cycle have a profound effect on the evolution of the chemical composition of the atmosphere and others influence the CO_2 changes only slightly. The CO_2 cycle resulting from the atmosphere-ocean gas exchange proceeds comparatively fast because of the difference between oceanic water temperatures in high and low latitudes and due to seasonal temperature variations.

At present the average concentration of atmospheric CO_2 is equal to about 0.034. Observational data show that the CO_2 concentration near the equator is somewhat higher than at high latitudes. This difference can be attributed to the greater solubility of carbon dioxide in cold ocean waters at high latitude than in the warm waters of the tropics. As a result, in high latitudes the atmosphere loses some CO_2 that dissolves in the oceans, where the excess of carbon dioxide is transferred into lower latitudes by cold deep currents and then returns into the atmosphere. The corresponding CO_2 flux between the equator and the North Pole is about 2×10^{16} g yr^{-1} (Bolin and Keeling 1963). The CO_2 exchange between atmosphere and ocean when the mixed layer temperature varies throughout the seasons is caused by a similar mechanism, the residence time of CO_2 molecules in the atmosphere being several years. In this case the process of gas exchange is practically closed (i.e. the annual amount of carbon dioxide received from and given to the ocean is the same). Therefore it does not significantly influence the CO_2 content of the atmosphere over long periods of time.

The biotic cycle of carbon dioxide, when carbon dioxide of the atmosphere and the hydrosphere is assimilated by autotrophic plants by photosynthesis is of great importance in maintaining life on Earth. The simplified photosynthetic reaction can be expressed as

$$CO_2 + H_2O = CH_2O + O_2. \tag{3}$$

This formula shows that in photosynthesis carbon dioxide and water yield carbohydrates and oxygen. In photosynthesis land plants utilize mainly atmospheric carbon dioxide and to a considerably less extent CO_2 from the soil. Aquatic plants absorb carbon dioxide dissolved in the waters of the hydrosphere. An appreciable part of photosynthetic organic matter is used for plant respiration and the atrophy of their organs, which returns carbon dioxide into the environment. Hence, autotrophic plant productivity and the total loss of carbon dioxide usually make up half or two thirds of the primary expenditure of CO_2 in photosynthesis.

Yefimova (1977), as well as several other authors, has calculated the value of autotrophic plant productivity on the continents and constructed a world map of photosynthetic productivity. She obtained a value for autotrophic plant productivity of 14×10^{16}g yr^{-1}. Estimates of oceanic plant productivity are less accurate. According to them, this productivity is assumed to be about 5×10^{16}g yr^{-1} (Koblenz-Mishke and Sorokin 1962). From these, the total autotrophic plant productivity would be 19×10^{16}g yr^{-1} of dry organic matter.

It is easy to calculate that with this level of productivity the atmosphere annually loses 28×10^{16}g of carbon dioxide. This corresponds to the average residence time of CO_2 molecules in the atmosphere of about 10 years, without taking the participation of hydrospheric carbon dioxide in the biotic cycle into account. The major portion of carbon dioxide assimilated by autotrophic plants returns into the atmosphere and the hydrosphere in organic matter destroyed by bacteria and other heterotrophic organisms. A comparatively small portion of photosynthetic organic carbon is retained for a long time in soil humus, sapropel, peat and other organic substances which are partially oxidized and partially deposited in the lithosphere as coal, petroleum, natural gases and dispersed organic carbon. As mentioned in Chapter 2, the rate of organic carbon deposit during different geological periods of the past can be derived from geochemical studies of the sedimentary shell. It is difficult to determine this rate for the present epoch, but it can be assumed to be approximately the same as in the late Neogene (the Pliocene), i.e. 0.3×10^{14}g yr^{-1}, as can be concluded from Chapter 2.2. The amount of carbon dioxide in modern atmosphere could therefore be exhausted within 10^5 years. Although this estimate increases considerably on taking the reserve of hydrospheric carbon dioxide into account, nevertheless it is obvious that without constant CO_2 influx from the depths of the Earth, the maintenance of an invariable atmospheric CO_2 concentration over a period comparable with past geological epochs is impossible.

A much greater expenditure of atmospheric and hydrospheric carbon dioxide is associated with carbonate formation. Carbonates are deposited more actively in shallow water bodies where the erosion products, including various carbonate combinations, are washed out from the continental surface. These erosion products are deposited as limestone, chalk, dolomite and other minerals containing carbon. Aquatic organisms play an important role in depositing carbonates because their skeletons consist of carbon compounds contained in water. Deposition of shells and other remnants of aquatic organisms make up a considerable portion of carbonate rocks.

Data in Chapter 2.2 show that the rate of CO_2 expenditure for depositing carbonate rocks in the Pliocene was 1.8×10^{14}g yr^{-1}, i.e. six times greater than that for depositing organic carbon. In Chapter 2.3 it is noted that the loss of CO_2 in carbonate formation at present appears to be somewhat less than in the Pliocene. The same conclusion can probably be drawn concerning the rate of CO_2 loss in the deposition of organic carbon.

Carbon dioxide enters the atmosphere from the depths of the Earth as a result of the degassing of the upper mantle and the higher layers of the crust.

A considerable quantity of atmospheric CO_2 is ejected during volcanic eruptions. It has long been known that carbon dioxide, together with water vapour, are the most important components of volcanic gases.

In addition to CO_2, volcanic gases usually contain a certain amount of carbon monoxide (CO) and methane (CH_4), which oxidize to carbon dioxide.

Other natural sources of atmospheric carbon dioxide are less important than the degassing of the mantle. In the anaerobic destruction of organic matter (in swamps, on rice fields and in termitaries), a certain amount of methane is formed which, on being oxidized, is converted into CO and then into CO_2. This process should be considered as a part of the total destruction of organic matter rather than an external (relative to the biosphere) carbon dioxide source.

There are no exact estimates of the rate of CO_2 production into the atmosphere from the depths of the Earth. It follows from the calculations of the atmospheric carbon dioxide balance described in Chapter 2.3 that over quite long periods (of the order of a million years or more), the rate of CO_2 gain in the atmosphere has been approximately equal to the loss rate. Thus, the rate of CO_2 production for the last several million years has been approximately equivalent to the rate of CO_2 expenditure.

However, it should be kept in mind that the rate of CO_2 production, both for long and short periods of time, can vary widely, mainly due to variations in volcanic intensity. Changes in atmospheric carbon dioxide content are thus unavoidable, as is discussed in detail in Chapter 3.1.

In the present epoch, the burning of coal, petroleum and other kinds of carbon fuel is an additional source of CO_2 in the atmosphere. Thus, in recent years about 5×10^{15} g of carbon enter the atmosphere annually, which much exceeds the amount of carbon received by the atmosphere from the depths of the Earth.

Carbon dioxide production by man is leading to noticeable changes in the chemical composition of the atmosphere, which are considered in detail in Chapter 3.

Oxygen. As noted above, atmospheric oxygen is largely produced in the photosynthesis of autotrophic plants. Taking the indicated value of the global photosynthetic productivity, we find that annually 20×10^{16} g of oxygen enter the atmosphere. Since the atmosphere contains 1.2×10^{21} g of oxygen, the duration of the oxygen cycle in the atmosphere due to photosynthesis is about 6000 years.

A certain amount of oxygen goes into the atmosphere as a result of water vapour photodissociation. When water vapour molecules are influenced by ultraviolet radiation in the 0.175-0.203 μm wavelength band, free hydrogen forms, with atoms in the upper layers of the atmosphere developing considerable velocity to overcome the Earth's gravity. The loss of a certain amount of hydrogen formed in water molecule dissociation results in the generation of an equivalent amount of free oxygen. The rate of this generation is difficult to estimate. Berkner and Marshall (1965a, 1966) found that without photosynthesis, water vapour photodissociation can produce an amount of oxygen not exceeding 0.1% of its modern content. In this case it is important to take into account the "Urey effect" (when a very small amount of oxygen is sufficient to absorb the part of the ultraviolet radiation spectrum very active in water dissociation). In this connection the screening effect of oxygen makes accumulation of its considerable mass impossible.

Some authors (Byutner 1961; Brinkman 1969) have obtained higher estimates for the effect of water vapour photodissociation as compared with the results for photosynthesis. The majority of modern reviews contend that photodissociation can maintain only a very small oxygen concentration, since the nature of the biological and geological processes before the wide dissemination of photosynthesizing plants indicates that the ancient atmosphere contained hardly any oxygen.

However, this inference is not very accurate, since in the Early Precambrian the atmosphere received a considerable amount of not fully oxidized gases which could absorb an appreciable amount of the oxygen produced in water vapour molecule photodissociation. Nevertheless, given the great amount of oxygen absorbing ultraviolet radiation in the present atmosphere, it is unlikely that quantities of oxygen comparable with those coming from autotrophic plant photosynthesis were produced in photodissociation.

The estimate of oxygen production is the difference between the rate of oxygen formation in photosynthesis and its loss for respiration and other vital functions of autotrophic plants. However, these types of O_2 consumption are not the only ones in its biotic cycle. The main portion of incoming oxygen is consumed by oxidizing organic matter of heterotrophic organisms and by the destruction of dead organic matter.

The total oxygen income in its biotic cycle can be estimated from the rate of organic carbon deposit in the late Neogene. As noted above, in the Pliocene 0.3×10^{14} g of organic carbon were deposited annually. This is equivalent to an annual oxygen gain in the atmosphere of 0.8×10^{14} g, the latter being only about 0.04% of photosynthetic oxygen.

The atmospheric oxygen formed in the biotic cycle is expended for oxidizing rocks and gases coming into the atmosphere from the depths of the Earth such as CO, SO_2, H_2S, H_2, etc.

A comparison of the two ways of consumption of biotic oxygen indicated has not yet been made. Data in Chapters 2 and 3 show that oxygen consumption on oxidizing these gases makes up less than a quarter of the oxygen production, while the oxygen amount used by oxidation of rocks is more than three fourths.

In addition, oxygen, like carbon dioxide, is absorbed in oceans and passes into the atmosphere from the oceans according to the balance of organic matter produced by aquatic autotrophic plants, and because of the dependence of oxygen solubility on water temperature. Unlike CO_2, the relative volume of O_2 dissolved in water bodies is noticeably less than the quantity of this gas in the atmosphere. The balance of oxygen in its exchange between the atmosphere and the ocean is almost completely closed similarly to the balance of carbon dioxide, therefore the influence of this exchange on the evolution of atmospheric chemical composition is limited.

In the modern epoch, a certain amount of oxygen is consumed in burning different kinds of carbon fuel. The rate of this loss, determined by the above value for annual consumption of carbon fuel, amounts to about 13×10^{15} g yr^{-1}. This value forms a considerable part of the rate of oxygen production in photosynthesis, exceeding that to the atmosphere by a factor of 100 to 200.

The question arises whether the burning of fossil fuel can lead to a considerable reduction in the quantity of oxygen in the atmosphere? Such a possibility obviously does not exist for the near future. At the present rate of anthropogenic consumption of oxygen, its amount decreases annually by 10 ppm of total mass of O_2 in the atmosphere. It is unlikely that in the future the carbon fuel consumption rate will increase greatly because the resources of this fuel are limited.

At the same time it can be found that even if the available resources of carbon fuel were fully exhausted, the amount of atmospheric oxygen would be reduced by only a fraction of a per cent of the atmosphere's volume.

Thus, man's impact on atmospheric oxygen content is incomparably less than on the carbon dioxide concentration, because the mass of atmospheric oxygen is approximately 500 times greater than that of carbon dioxide.

In summary, we note that the rate of atmospheric oxygen consumption depends in part on its partial pressure (e.g. in oxidizing rocks) and is determined partly by external factors (in particular, by the rate of production of not fully oxidized volcanic gases). However, in all cases this rate is independent of the rate of oxygen income, which makes considerable changes possible in content of atmospheric oxygen with time. These changes are restricted to some extent by negative feedbacks between oxygen concentration and (a) its loss in oxidizing minerals; (b) the rate of organic carbon deposition (which decreases with increasing oxygen concentration).

Note that with the great volume of atmospheric O_2, even noticeable differences in the rate of gain and loss of oxygen can only result in considerable changes in this volume over long time periods of millions of years.

Nitrogen. The nitrogen cycle greatly differs from that of carbon dioxide and oxygen. This can be attributed largely to the smaller chemical activity of nitrogen as compared with the indicated gases.

Nitrogen, like carbon dioxide, enters the atmosphere from the depth of the Earth in the degassing of the mantle. However, the amount of volcanic nitrogen is noticeably less than that of CO_2. At the same time, the rate of depositing atmospheric nitrogen is much less than that of atmospheric carbon. Deposits contain about 1.0×10^{21} g of nitrogen (Garrels et al. 1975). Assuming that nitrogen deposited largely in the Phanerozoic, we find that the average rate of deposition over the indicated period of time was 1.7×10^{18} g (m.y.)$^{-1}$. This value is less than the rate of carbon deposition by a factor of approximately 100.

As can be seen from the calculations in Chapter 3.1, throughout the Phanerozoic the nitrogen content of the atmosphere varied only comparatively little. Thus the rate of nitrogen production to the atmosphere averaged over this time period was close to the rate of that to the lithosphere. Therefore we can conclude that the amount of nitrogen in gases released from the depths of the Earth into the atmosphere made up about 1% of the carbon content of these gases.

It follows from these information that the duration of the nitrogen cycle between the atmosphere and the lithosphere is very long, approximately 2.0

billion years. This explains why, in the absence of a direct relationship between gain and loss rates of atmospheric nitrogen in this cycle, its quantity over the last billion years could not have changed considerably. Before this period the situation was different.

In addition to nitrogen exchange in the system atmosphere-Earth's depths, an internal nitrogen cycle exists involving the atmosphere, the soil and, to some extent, the hydrosphere. This cycle, including a comparatively small portion of atmospheric nitrogen, is practically closed. Thus, in the internal cycle there is no considerable increase or decrease in the total amount of free nitrogen. Since this cycle does not directly affect the evolution of atmospheric chemical composition, we consider it here only briefly.

As noted above, due to the activity of soil and aquatic nitrogen-fixing bacteria and algae, ammonia and other nitrogen compounds form from atmospheric nitrogen. These compounds are partially oxidized by free oxygen, partially retained in soil and water, or assimilated by plants. Plant nitrogen is transferred to heterotrophic organisms by trophic links.

Nitrogen compounds in soil, water and in the remnants of organisms after denitrification produce free nitrogen that returns into the atmosphere. Denitrifying bacteria inhabiting soil and water that contain insignificant amounts of oxygen are important in this process. Due to their activity, both free nitrogen and a small quantity of N_2O are produced. N_2O goes into the stratosphere, where through photochemical reactions it is converted into NO and nitrogen. NO interacts with stratospheric ozone and oxidizes to NO_2, thus decreasing the ozone content.

The internal cycle of nitrogen is not very intense, therefore the results of economic activities (production of nitrogen fertilizers, etc.) often prove to be comparable with the natural nitrogen cycle. Since the atmospheric nitrogen volume is great, anthropogenic factors hardly influence it. The internal nitrogen cycle can exert an indirect influence on atmospheric evolution because of the necessity in photosynthesis for plants to use nitrogen compounds whose content in soil and hydrospheric waters is often insufficient to maintain high photosynthetic productivity. The lack of phosphorus compounds and other minerals necessary for photosynthesis has a similar effect on plant productivity in many regions. Limited plant productivity results in decreasing the rate of gain in atmospheric oxygen and in increasing, other things being equal, the atmospheric carbon dioxide concentration.

1.3 Studies of the Evolution of the Atmosphere

Problems in Studying the History of the Atmosphere. The question of changes in the atmosphere in the geological past is part of the more general problem of studying the Earth's evolution. The history of the individual components of the Earth's external envelope has only been studied sporadically and the amount of data available on the evolution of each major component varies.

The abundant material on the rocks formed during different geological epochs has led to important results for studying the history of the lithosphere, which is of great practical value.

The great progress in studies of natural history has been due both to theoretical consideration of organic evolution and to the presence of extensive empirical material on plants and animals in the past. It is natural that the sciences that study the history of the lithosphere and of organisms, historical geology and palaentology, should have emerged long ago. These two disciplines play a significant role in Earth Sciences. The history of the other two components of the Earth's external envelope, the hydrosphere and the atmosphere, has been much less studied. The publications devoted to the hydrosphere and the atmosphere are much fewer than those on the history of the lithosphere and organisms. It is no wonder therefore that special sciences treating the history of the hydrosphere and the atmosphere have not yet emerged. Although the term "palaeoceanology" is sometimes used in works on the history of the hydrosphere, it refers rather to the oceans than to the hydrosphere as a whole and has not been universally adopted. A name for the science concerning the history of the atmosphere has not yet been proposed. It is obvious that the term "palaeoclimatology" cannot be used in this case, since in palaeoclimatic studies only climates of the past are investigated and, as a rule, the evolution of atmospheric chemical composition, that is a central question in the history of the atmosphere, is not touched upon at all.

Without discussing the history of the hydrosphere, which is beyond the framework of this book, let us note that the very small number of studies concerned with the history of the atmosphere that have been carried out previously can be explained by two reasons. The first is the absence of easily interpreted empirical materials on the atmospheric composition in the geological past. The second is the underestimation of the importance of the history of the atmosphere in the study of the evolution of the Earth as a whole.

The methods available for studying the history of the atmosphere are considered in the subsequent paragraphs of this section. Here we only note that the possibilities of studying the history of the atmosphere are widened considerably by using quantitative methods for analyzing information on the Earth's sedimentary shell.

The question of the importance of data on the history of the atmosphere in studying the evolution of the Earth should be discussed in more detail. There is no doubt that the evolution of the Earth's external envelope was complex. Therefore, in order to study the features of this process, it is necessary to know the mechanism of interactions between the major time-dependent components of the external envelope, including the atmosphere. The consideration of the effects of atmospheric factors in the study of the lithogenesis and evolution of organisms is particularly important. Some conclusions referring to the latter problem are presented in Chapter 3.

An important step in studying the Earth's history was to overcome the anti-evolutionary concepts particularly widespread in biology. This took a considerable time; attempts have been made even recently to narrow the sense of the modern theory of the evolution of organisms or even to reject entirely the idea of animal and plant evolution (creationism).

Uniformitarianism, i.e. an extreme form of actualism, was an analogue of the anti-evolutionary biological concepts in geology. As is known, the principle of actualism, stated by Lyell in the first half of the 19th century, was based on the assumption that in the past the Earth's crust had changed as it does at present. Although this approach allowed us to understand many of the changes in the lithosphere, it was later established that a number of events in the history of the lithosphere cannot be explained by modern lithogenesis alone. Since in the past this process has changed qualitatively, it is obvious that the actualistic approach in studying the history of the Earth's crust is limited. Although anti-evolutionary ideas in biological and geological sciences are less widespread now, in investigations of the history of the atmosphere, similar concepts often occur.

The conclusion that in the remote past the chemical composition of the atmosphere differed only slightly from the modern one was usually corroborated by the similarity in the manner of formation of various minerals, depending on the atmospheric chemical composition during geological epochs, including ancient ones. These assertions were, as a rule, arbitrary and did not assist in clarifying the real history of the atmosphere. More prudent supporters of the theory of the invariability of atmospheric composition limited the period of existence of the modern atmosphere to only a part of its entire history, usually to the Phanerozoic. This point of view was also arbitrary.

There are less objections to the idea that the composition of the atmosphere could only change within a limited range because the balance between major atmospheric gases and their concentrations stabilizes the atmospheric composition. While admitting the importance of this balance, we should emphasize that it cannot exclude considerable variations in the amounts of individual gases of which the atmosphere is composed. Meanwhile negative feedbacks between concentrations of atmospheric gases and their ratios are sometimes considered as proof of the invariability of the composition of the atmosphere.

The fact that anti-evolutionary views on the history of the atmosphere are still to be found can be attributed to the difficulties in studying the chemical composition of the atmosphere in the geological past and, particularly, in obtaining valid quantitative information on these changes. In this case it is possible that the principle of "economy of thought" operates, i.e. given a complicated scientific problem, the simplest way to solve it is to say that it does not exist at all.

It should, however, be emphasized that even with the lack of valid data on the chemical composition of the atmosphere in the geological past, it is hardly possible to believe that this composition has been constant over billions of years.

It has long been known that autotrophic plants considerably influence the balance of two atmospheric gases, CO_2 and O_2. At the beginning of Earth's history, photosynthesizing plants did not exist. Then they appeared and, progressively changing, undoubtedly produced an important effect on atmospheric composition.

Several decades ago it was found that in the early epochs of the Earth's history the luminosity of the Sun was appreciably lower than at present. In order to explain the existence of warm climatic conditions in the remote past, it is necessary to assume that the atmosphere at that time differed in chemical composition from the present one.

Not to mention other simple ideas that clearly contradict the theory of the invariability of the atmosphere, we note that in recent times, after quantitative calculations of changes in atmospheric composition, the previously widespread opinions as to the invariability of the atmosphere occur much more rarely.

Returning again to the question of the importance of data on the Earth's history in studying its evolution, we note first that these data are of methodological importance since they demonstrate the atmosphere's participation in a general evolutionary process embracing the Earth's external envelopes.

Second, as noted above, many features of the complex process of the Earth's evolution can be understood only by taking into account changes in atmospheric composition which on the one hand resulted from the processes occurring in the depths of the Earth, the upper layers of the lithosphere and in the biosphere, and on the other hand had an appreciable effect on the lithosphere and the biosphere.

Third, the question of atmospheric evolution has recently acquired considerable practical importance, due to the present discovery of the process of anthropogenic growth of atmospheric carbon dioxide concentration. The possibility has thus emerged of using data on the chemical composition of the atmosphere in the past to estimate climatic conditions of the near future. This problem is discussed in detail in Chapter 3.1.

Qualitative Characteristics of Atmospheric Composition. Reviewing the most important studies of the history of the atmosphere, we can divide them into two groups, the first describing qualitative and the second quantitative estimates of changes in the atmosphere. This division is to some extent conventional. In particular, several studies from the first group present quantitative characteristics of the chemical composition of the past. However, these estimates are either based on intuitive reasoning rather than on well-founded calculation, or contain limited information on atmospheric components (determination of the order of magnitude of the components' concentration, inferences as to lower or upper limit of their values, etc.). At the same time, the use of quantitative methods of study with the latest available knowledge does not always result in obtaining accurate information.

We think that in spite of some reserves, this division is useful, since it shows a certain change in the methods of studying the history of the atmosphere that occurred in the 1970's.

The idea that in the geological past the atmosphere changed considerably appeared very long ago. In particular, it was proposed in the first half of the last century by the eminent French biologist Geoffroy St. Hilaire (1772-1844), who was the first to show the importance of changes in the atmosphere for organic evolution.

Later, Tyndall (1861) noted that since carbon dioxide, together with water vapour, absorbs outgoing longwave radiation in the atmosphere, changes in carbon dioxide concentration could have been one of the causes that led to changes in climatic conditions in the past.

Tyndall's idea was later developed in studies by Arrhenius (1896, 1903) and Chamberlin (1897, 1898, 1899), who assumed that changes in CO_2 content could have caused the Quaternary glaciations.

Taking into account the results from geological studies, Arrhenius noted that the modern atmosphere contained a small amount of the carbon dioxide which in the past had been consumed from the atmosphere for carbonate formation. Therefore, Arrhenius concluded that atmospheric CO_2 concentration could vary widely. These changes considerably affected air temperature, which in turn could cause glaciations to advance and to retreat.

Studying the CO_2 balance in the atmosphere, Chamberlin concluded that the production of carbon dioxide from the lithosphere varied considerably, depending on the level of volcanic activity and other factors. The expenditure of carbon dioxide also varied greatly, in particular because of differences in the area of the rocky surface subjected to atmospheric erosion. With increase of this area, CO_2 loss caused by weathering rose. Chamberlin assumed that glaciations advanced as a result of an intense orogenic process and emergence of the continents, which led to increasing erosion depth, growth in weathering rock area and reduction in atmospheric CO_2 concentration. To corroborate this assumption, Chamberlin carried out calculations, which, however, cannot be considered as a satisfactory quantitative model of the process under consideration.

Later the assumption of effects of CO_2 concentration on climate was doubted. It was found that in the 13-17 μm wavelength band of radiation absorption by carbon dioxide water vapour absorbs radiation, which decreases the temperature effects of the change in CO_2. The conclusion was drawn that the cause of the Quaternary glaciations is not associated with changes in atmospheric chemical composition.

Recent studies of the chemical composition of air bubbles in ancient ice showed that during glaciation noticeable changes in atmospheric CO_2 concentration occurred. It decreased to approximately 0.02%. At the same time it has been found that taking the effects of carbon dioxide and water vapour on the absorption of longwave radiation correctly into account, the dependence of air temperature on CO_2 concentration is important and corresponds approximately to the relationship obtained by Arrhenius (see Chap. 3). These results have corroborated the forgotten hypotheses of Arrhenius and Chamberlin (Manabe and Broccoly 1985).

However, they have not disproved the basic statements of Milankovich's theory, which assumed that glaciations advanced due to periodic oscillations of the position of the Earth's surface relative to the Sun. The appearance of the Quaternary glaciations seems to be associated with changes in astronomical factors studied by Milankovich. Their climatic effects increased due to variations in carbon dioxide concentration.

The ideas of Arrhenius described in his book (1908), published soon after the appearance of his studies on the Quaternary glaciations, are of great interest. Arrhenius emphasized that, within wide limits, a close relationship existed between carbon dioxide concentration and volcanic activity level. He assumed that volcanic activity increased in the Eocene and the Miocene, causing a rise in CO_2 concentration and in mean air temperature. At the same time photosynthetic productivity grew. Arrhenius believed that in the Eocene, due to the strengthening greenhouse effect, the mean air temperature was $8°-9°C$ higher than the modern one. He attributed the strong development of continental vegetation in the Carboniferous period to increased CO_2 concentration, which in turn resulted in an increasing amount of atmospheric oxygen.

Note that all these conclusions of Arrhenius coincide with the results presented in this book. The coincidence of estimates of mean air temperature for the Eocene, which, according to our calculations, is $8.5°C$ higher than at present (Table 10), is striking. In this case the similarity is to a considerable extent accidental, since the two calculations differed somewhat. However, analyzing these inferences of Arrhenius, we can conclude that on the whole they were much ahead of the ideas of atmospheric science at the beginning of our century.

Among other studies on the evolution of the atmosphere, one by Grigoriev (1936) is of specific interest, he assumed that the amounts of carbon dioxide and oxygen in the past atmosphere were different, variations in carbon dioxide content being greater than those in oxygen concentration. He also believed that by increasing the amount of carbon dioxide produced from the Earth's depths, the growth in atmospheric CO_2 concentration could not have been considerably decreased by a rise in organic biomass or absorption of carbon dioxide by the ocean. In this work it was also noted that climatic conditions in the past were affected by a change in carbon dioxide concentration, and the question was raised as to the possible importance of variations in the atmospheric oxygen content for the evolution of organisms.

Berkner and Marshall (1965a, 1965b, 1966) endeavoured to construct a graph of changes in atmospheric oxygen content throughout the Phanerozoic. In these studies they came to the conclusion that, in the absence of photosynthesis, only a small quantity of atmospheric oxygen can form in the photodissociation of water vapour. In constructing the curve of changes in oxygen concentration during the Phanerozoic, Berkner and Marshall tried to relate biological processes to oxygen content, which appeared to be a complicated problem. They thought that multicellular organisms emerged when the mass of oxygen was more than 1% of its modern value, and life disseminated over land after the formation of the ozone layer when the oxygen mass was 10% of its modern value. Assuming that the former event occurred at the beginning of the Cambrian and the latter in the Devonian, and using hypothetical data on variations in oxygen amount over other periods of time, Berkner and Marshall modelled the curves of variations in oxygen during the Phanerozoic.

Such an approach to this studying raises objections. Firstly, the idea of Berkner and Marshall that in the Precambrian multicellular organisms did not exist was incorrect. Secondly, it is not obvious that these organisms could appear when the amount of oxygen in the atmosphere was 1% of its modern volume. A similar objection can be expressed to the second hypothesis of these authors about the emergence of organisms on land when the volume of oxygen was 10% of its modern value.

Finally it is not clear how it was possible, even if these hypotheses were correct, to construct a complicated curve characterizing changes in the mass of oxygen during the Phanerozoic by using only three points in the graph (the third referred to the modern oxygen content).

Comparing the curves of Berkner and Marshall with our results shown in Chapter 3, we can draw an interesting conclusion. Although the above-mentioned two hypotheses of these authors appeared to be very inaccurate (particularly the first one of the much underestimated oxygen volume in the early Cambrian), the curves compared (in particular, the curve from their second work) proved to be rather similar to the curves calculated in Chapter 3. This similarity demonstrates that certain relationships exist between the history of organisms, taken into account by Berkner and Marshall, and variations in oxygen volume. However, this method of study allowed us to find only some qualitative characteristics of changes in oxygen mass and appeared to be less effective in obtaining quantitative estimates.

It was no chance that a detailed discussion of the evolution of the oxygen in the atmosphere in the review by Van Valen (1971) some years after the publication of the studies of Berkner and Marshall begins by questioning the validity of the latter's results. Van Valen, having considered the major features of the atmospheric oxygen balance, pointed out the basic relations between its components, which limit variations in the volume of atmospheric oxygen, and discussed the unsolved questions concerning variations in this volume. In summary, Van Valen concluded that quantitatively the problem of variations in oxygen concentration remained unsolved.

The idea of estimating variations in atmospheric content from data on changes in lithogenesis was proposed later by Sochava and Glickman (1973) and Sochava (1979). The value of these studies lies in an analysis of feedback between oxygen and carbon dioxide balance and the concentrations of these gases. The main qualitative results of investigations of the evolution of the atmosphere are described in a number of reviews published by well-known specialists in the field of geology and geochemistry, who themselves contributed greatly to studying the evolution of the atmosphere. Among these reviews are those by Rutten (1971), Cloud (1974) and Holland (1984).

As can be seen, Rutten, in discussing the problem of the evolution of the atmosphere, was influenced by the ideas of Berkner and Marshall, although in some questions he entertained his own opinion, differing from theirs.

Rutten assumed that the early secondary atmosphere contained about 0.1% of the modern oxygen volume, oxygen being formed at that time in water molecule photodissociation. About three billion years ago, the oxygen content increased due to photosynthesis, before 1.8 billion years ago the mass of oxygen did not exceed 1% of its modern value, which corresponded to Rutten's concept of the atmosphere without oxygen. Then the amount of oxygen began growing rapidly. By the Cambrian it had reached several per cent of its modern value (in this question Rutten partly corrected the explicit error of Berkner and Marshall). In the Silurian the relative amount of oxygen was 10%, which made the dissemination of life over land possible. Later, the amount of oxygen varied, decreasing during the periods of orogenesis and growth of volcanic activity, and increasing between these periods because of the gradual rise in the rate of photosynthesis. Rutten believed that in the Carboniferous the amount of oxygen was maximal, somewhat exceeding the present atmospheric oxygen content.

As to carbon dioxide, Rutten assumed that beginning with the Early Precambrian its volume gradually decreased, the maxima and minima of concentrations being observed in the epochs with increased and reduced volcanic activity. Rutten thought that maximum atmospheric CO_2 concentration in the Early Precambrian was ten fold greater than its modern value. In the Phanerozoic the amount of carbon dioxide, according to Rutten, varied from a value exceeding its modern level by several times up to a value somewhat smaller than the modern one.

Rutten made wider use of hypotheses than the authors of the reviews concerning the evolution of the atmosphere considered below. In this respect his approach resembles the treatment of these questions by Berkner and Marshall.

Comparing Rutten's assumptions about variations in atmospheric CO_2 and O_2 content with the calculated results given in Chapter 3, it can be noticed that even if Rutten's estimates were very approximate, his ideas about qualitative features of variations in atmospheric content were correct.

Conclusions about the evolution of the atmosphere drawn by Cloud (1974) in his review in the Encyclopaedia Britannica are in better agreement with calculated results given in this book. Noting that one can only guess about the content of the primary atmosphere, Cloud indicated the possibility of studying the history of the atmosphere, beginning 3.6 billion years ago. From this time up to 1.9 billion years ago the atmosphere did not contain oxygen and consisted largely of water vapour and carbon dioxide, including some quantities of CO, CH_4, NH_3, etc. Oxygen, produced at this time in water vapour photodissociation and then through photosynthesis, was completely consumed in interacting with not fully oxidized atmospheric gases and rocks.

The fact that the ancient atmosphere did not contain oxygen is explained, according to Cloud, by several causes: (1) chemical evolution resulting in the creation of living matter is possible only without oxygen; (2) deposits aged over two billion years contain iron and uranium compounds that could not have been formed in an oxygen atmosphere; (3) these deposits contain very small amounts of compounds of different chemicals with oxygen.

In the range of 1.8 to 2.0 billion years ago, strongly oxidized iron compounds appeared, which indicated the presence of free oxygen in the atmosphere. This meant that the amount of oxygen produced in photosynthesis exceeded its losses in oxidizing rocks and gases produced from the depths of the Earth. By the Phanerozoic the amount of oxygen in the atmosphere, in Cloud's opinion, was several per cent of the modern content. Throughout the Phanerozoic the amount of atmospheric oxygen varied noticeably, depending on volcanic activity and some other factors. Although Cloud assumed that data on these variations would be obtained in further studies, he pointed out the possibility of decreasing oxygen content in the Late Permian-Triassic and in the Late Cretaceous. Cloud indicated the insufficient validity of the hypothesis that between the Silurian and the Devonian the amount of atmospheric oxygen made up 10% of its modern content.

In addition to interpreting the history of atmospheric oxygen, Cloud pointed out an increase in the amount of atmospheric nitrogen occurring over billions of years, and the tendency of the atmospheric carbon dioxide concentration to decrease.

It is interesting that there is no essential contradiction between Cloud's views and the results obtained by the authors of this book in quantitative studies. However, Cloud, correctly pointing out the invalidity of both estimates of oxygen content during the "critical epochs" described by Berkner and Marshall, cited a somewhat underestimated value of oxygen mass for the first of these "critical epochs", i.e. for the onset of the Cambrian period.

The book by Holland was published 10 years after the review by Cloud. Almost the entire part of the book concerning the atmosphere contains the results of various geochemical studies, treating a number of processes occurring in the atmosphere. At the same time there is rather limited quantitative information on the evolution of the atmosphere. Only the last pages of the book contain some qualitative inferences as to changes in the chemical composition of the atmosphere in the geological past.

Noting that 2.0-3.0 billion years ago the amount of atmospheric oxygen was insignificant, the author assumed that an increase in the mass of oxygen occurred possibly 2.0-2.3 rather than 1.8-1.9 billion years ago, as some authors believed earlier.

Since the author failed to draw definite conclusions about changes in the chemical composition of the atmosphere during the Phanerozoic from geochemical studies, he used biological analogies and resorted to hypothetical ideas. The results of this approach appeared to be rather restricted. Among them is the conclusion that early in the Cambrian the volume of oxygen in the atmosphere was no less than 0.1 of its modern quantity, the upper limit of this value remaining unknown.

In addition, Holland assumed that in the Devonian, when plants spread over the land, the oxygen content increased and the carbon dioxide concentration in the atmosphere decreased. Then a new increase in oxygen volume occurred in the Cretaceous, and in the Cenozoic the amount of oxygen was similar to the modern one. At the same time Holland believed that during the Phanerozoic the volume of atmospheric nitrogen did not change considerably.

The majority of Holland's assumptions agree with the calculated results given in Chapter 3, and only two of them seem to be inaccurate. Thus, calculations show that the most considerable increase in oxygen concentration took place in the Jurassic and not in the Cretaceous period. The hypothesis that CO_2 volume decreased in the Devonian was not corroborated. At that time, as can be seen from data in Chapter 3, the rate of CO_2 income from the Earth's depths to the atmosphere increased rapidly, reaching a maximum in the Early Carboniferous. Therefore the CO_2 ·mass did not decrease in the Devonian, but considerably increased.

In addition to the studies indicated, some estimations of chemical composition of the atmosphere were carried out by Tappan (1968, 1970) and some other authors. The estimates obtained reflect correctly individual qualitative features in the composition variations; however, they do not yield accurate quantitative results. In conclusion, let us characterize briefly the modern state of investigations of the evolution of the atmosphere. The history of the atmosphere can be divided into three stages, according to the volume of material available on its past. The first stage is the beginning of the Precambrian, when the primary atmosphere existed and the secondary one was beginning to form. There are no valid empirical data on chemical composition and physical state of the atmosphere for this time, therefore studies of the atmosphere of this epoch are based on assumptions that are difficult to check.

For example, we cite briefly the description of the atmosphere at the beginning of the Precambrian by Walker (1976). He supports the conclusion that the Earth's primary atmosphere formed from a gas-dust cloud representing the source of matter for the construction of the Solar System. Walker believed that the volume of the primary atmosphere was insignificant, and that this atmosphere disappeared before the secondary atmosphere emerged.

The secondary atmosphere consisted of gases produced by the degassing of the upper mantle and crust of the Earth. Walker agrees with the widespread opinion that the early secondary atmosphere was not highly deoxidizing, although it contained hardly any free oxygen. This atmosphere consisted mainly of carbon dioxide and water vapour. Partial pressure of carbon dioxide in the early secondary atmosphere could make up about 5% of modern pressure, which is equivalent to a value of atmospheric CO_2 concentration exceeding the modern one by a factor of 100. This great volume of carbon dioxide caused a high rate of carbonate formation.

The early secondary atmosphere contained a relatively small mass of nitrogen and hydrogen, the quantity of the latter being determined by the balance between its production during a higher level of volcanic activity (compared to the present) and loss of molecules going into space. Due to the small quantity of hydrogen, the ancient atmosphere was weakly deoxidizing. Walker emphasizes that the assumptions on which this description is based are subject to many conditions.

The second stage in the history of the atmosphere includes a greater part of the Precambrian up to the onset of the Phanerozoic. For this time it is possible to investigate the past of the atmosphere empirically, although the information available is frequently insufficient to estimate accurately even the greatest changes in the state of the atmosphere. There is much more information for the third stage i.e. for the Phanerozoic, which makes it possible to investigate the atmosphere's evolution quantitatively at this very important stage.

It is interesting that most studies treat the first and the second stages indicated here. If it is difficult to understand the particular interest shown in the ancient history of the atmosphere, about which only more or less probable hypotheses can be advanced, it is even more difficult to explain the scarcity of attempts to clarify the features of the Phanerozoic history of the atmosphere, which has been covered by extensive data. This fact cannot be easily explained. Some authors also have doubts about the possibility of determining changes in the chemical composition of the atmosphere in the Phanerozoic, whereas for the Precambrian these changes are determined rather frequently, as a rule on the basis of very insufficient empirical data.

We think that without thorough studies of the evolution of the atmosphere in the Phanerozoic it is impossible to investigate the earlier history successfully. The state of available information, as we believe, reveals the main events of the evolution of the atmosphere in the Phanerozoic. This analysis can be supplemented with much more restricted and approximate conclusions about some features of the evolution of the atmosphere in the Precambrian.

Quantitative methods, it seems to us, look most promising for studying the atmosphere's history. These are discussed below.

Quantitative Characteristics of the State of the Atmosphere. Since the mid-1970's, the authors of this book have studied changes in the chemical composition of the atmosphere in the geological past by using quantitative calculations. Tentative results have been published in several works: Budyko (1977a, 1977c), Budyko and Ronov (1979), Budyko et al. (1985).

This study appeared possible due to information on global changes in the rate of carbon accumulation in deposits obtained by Ronov and his co-workers. The works where this information appeared are discussed in Sections 1 and 2 of Chapter 2.

In the above-mentioned publications, data on rates of carbonate rock and organic carbon accumulation have been used to determine changes in the amount of carbon dioxide and oxygen throughout the Phanerozoic. In preparing this book, the calculations have been improved and supplemented with estimates of changes in carbon dioxide and oxygen concentrations during the Precambrian and also in atmospheric nitrogen content. The results obtained are treated in Chapter 3.

In addition to these studies, other investigations have been carried out, aimed at determining the quantitative characteristics of the atmosphere in the geological past. Let us consider the most important of these investigations.

Several publications have dealt with the possibility of estimating by data on the past climatic conditions the amount of atmospheric gases producing the greenhouse effect. The hypothesis of Arrhenius mentioned in the previous section, that warmer climates of the past can be attributed to a higher concentration of atmospheric carbon dioxide, has been confirmed by one of the authors of this book (Budyko 1974). In his study, the conclusion was drawn that throughout the Mesozoic Era and over a greater part of the Tertiary, due to increased atmospheric CO_2 concentration, mean air temperature was higher compared to the present value by approximately 5°C.

Dobrodeev and Suetova (1976) have calculated carbon dioxide concentrations that could have caused the changes in temperature of the mixed ocean layer in different epochs of the Tertiary, discovered by Emiliani. These authors found that the CO_2 concentration exceeded the modern one by a factor of 16 in the Eocene, 8 in the Oligocene, 4 in the Miocene and 2 in the Early Pliocene. Comparing the results of this calculation with data in Table 8, we can see that the indicated results correctly reflect the trend to a decrease in atmospheric carbon dioxide throughout the Cenozoic. Greater CO_2 concentrations obtained in this calculation compared to values presented in Table 8 seem to be attributed to the fact that data on temperature changes in extratropical latitudes have been used. These changes, as is known from palaeoclimatic investigations, are noticeably greater than global mean temperature changes that should be taken into account in calculating carbon dioxide concentrations. The absence in this calculation of the inference as to declining CO_2 concentration in the Oligocene is the consequence of the insufficient representativeness of the data on palaeotemperatures in this epoch, since more recent information on climatic change allows us, as a rule, to discover considerable cooling in the Oligocene (see Chap. 3).

Similar calculations of carbon dioxide concentration in the geological past using data on palaeotemperatures have been carried out in the USSR and other countries. Data on variations in mean global palaeotemperatures have been used in these calculations, the dependence of surface air temperature on changes in the Sun's luminosity and the mean Earth's albedo being regarded.

By analogous calculations an inverse problem can be solved, i.e. checking the results of determining CO_2 concentration in geochemical studies from empirical evidence of past climatic conditions. This has been done in a number of studies, some results being presented in Chapter 3.1.

Hart (1978) has proposed another way to study the history of the atmosphere. He built a numerical model of the atmosphere's evolution that enabled him to calculate for its entire history both variations in chemical composition and in climatic conditions. The degassing rate, decreasing exponentially with a characteristic time of 8×10^8 yr, the composition of gases produced from the depths of the Earth (constant with time) and the solar radiation variations were chosen as external factors in his model. Hart's model incorporated very simplified equations describing an increase in ocean volume as atmospheric water vapour is condensed, photodissociation of water vapour molecules and hydrogen efflux into space, oxidation of surface rocks, loss of CO_2 on carbonate deposition, expenditure of CO_2 and H_2O in photosynthesis for the formation of organic matter and oxygen and sedimentation of organic carbon.

A system of equations describing climatic changes took into account albedo variations due to alterations of cloudiness and ice-cover areas and the influence on air temperature of the greenhouse effect caused by different gases. Several parameters of these equations, known only approximately or even completely unknown, were varied by Hart so that results could be obtained similar to the empirical conclusions drawn in studying the history of the atmosphere. Hart's study is of considerable importance because it appeared to be the first attempt to construct a more or less complete theoretical model of the evolution of not only the atmosphere but also the hydrosphere, as well as some components of the biosphere and lithosphere. It is less clear to what extent his inferences are realistic, since the theory of atmospheric evolution in this study has been considerably simplified.

The conclusion of Hart concerning the biosphere's high sensitivity to changes in external factors deserves attention.

Hart noted that a comparatively small decrease in solar radiation leads to the Earth's complete glaciation and a relatively small increase to the disappearance of liquid water. It is obvious that both these events imply the destruction of the biosphere. A similar conclusion has been drawn in several other modern investigations, which increases its validity.

The calculations of the evolution of atmospheric chemical composition of Hart are less reliable than his general conclusions. The following criticisms appeared soon after the publication of his study: (1) Hart's conclusion that a highly reducing atmosphere existed for the first two or three billion years of the atmosphere's history does not agree with the results of modern geochemical investigations; (2) the climatic part of Hart's model is imperfect, in particular the conclusion that mean global temperature is much affected by variations in cloudiness is highly improbable (Owen et al. 1979).

Concurring with these remarks, we can add some more ideas about the major reasons for the limited validity of the results obtained by Hart. Considering data presented in Chapter 3, we can conclude that decreasing the degassing rate was exponential only in the early stages of the history of the atmosphere; later it slowed down. An important feature of variations in the degassing rate was their short-term fluctuations expressed in data on sediments. Since Hart did not take these data into account, he could not obtain a correct pattern of variations in the chemical composition of the atmosphere during the Phanerozoic. Although a list of such remarks can be continued, they do not contradict the above conclusion that Hart's investigations are nevertheless important.

Owen et al. (1979) have calculated the variations in mean air temperature in the geological past due to increasing solar radiation and changes in CO_2 volume estimated by Hart. They used a more advanced climate model than that applied by Hart. Owen et al.'s calculation was performed to illustrate the possibility of obtaining realistic values of mean temperature without considering the influence on the greenhouse effect of other gases besides CO_2 and water vapour.

An interesting inference can be drawn from this calculation that since the time of 3.5 billion years ago up to the present epoch, the mean surface air temperature has varied within the very narrow range of 10°C.

Later, Kuhn and Kasting (1983) carried out a similar calculation. They supplemented the climate model used in the previous study with more detailed consideration of the atmospheric water cycle. This resulted in obtaining even smaller temperature variations throughout the entire history of the atmosphere.

The mechanism limiting air temperature variations, when external climate-forming factors change, has been discussed in several papers (see Budyko 1980).

The authors of this book in their above-mentioned studies have assumed that the CO_2 consumption rate is proportional to its concentration. Since air temperature increases as CO_2 concentration grows, it is clear that this assumption has included the effects of air temperature rise on the rate of CO_2 consumption in carbonate sedimentation.

Walker et al. (1981) related the indicated relationship to the effects of continental surface temperature on the weathering rate, whose increase yields an increase in carbonate sedimentation.

The relationship between the rate of silicate weathering (and, therefore, the rate of carbonate sedimentation) and the carbon dioxide concentration proved to be weaker than that used by the authors of this book in their studies. We assumed the rate of carbonate sedimentation to be proportional to the CO_2 partial pressure to the first power, while Walker et al. thought that it was proportional to the partial pressure to the power 0.3. At the same time, they believed that this rate was proportional to a mean value of river runoff increasing with rising air temperature, as follows from calculations carried out by using climate theory methods.

Taking into account a relationship of air temperature-carbon dioxide concentration, Walker et al. calculated the degassing rate necessary to save the climatic conditions similar to the contemporary, in the remote past when the Sun's luminosity was considerably less than at present.

As a result of considering the temperature-weathering rate feedback, the primary degassing rate appeared to be less than that obtained without due regard to this feedback effect.

We think that the relationship between CO_2 loss and concentration is not limited by the comparatively simple relations presented here. Some of the hypotheses accepted by Walker et al. seem to be inadequate. For instance, the relationship between air temperature and CO_2 concentration does not agree with the most valid results obtained by climate theory models. It is unclear whether it is possible to use the relationship between the global value of river runoff and mean air temperature determined for modern topography in calculations for the geological past when the position of the continents and the oceans (and therefore, of circulation zones) differed noticeably from today.

Without discussing other details of this study, we note the difficulties in fitting the hypothesis of the relationship between the rate of carbonate sedimentation and CO_2 concentration with empirical data on variations in carbonate sedimentation rate. As can be seen from data in Chapters 2 and 3, the carbonate sedimentation rate averaged over geological epochs varied during the Phanerozoic by a factor of 10. This, other things being equal, should have resulted in an approximately 2000–fold variation of CO_2 concentration if calculated by using Walker's et al. relationship. This variation is clearly incompatible with the survival of living organisms which existed throughout the Phanerozoic. Taking the effects on CO_2 concentration of possible runoff inconstancy into account does not seem to considerably decrease this variation.

Byutner's (1984) attempts to overcome the major shortcomings of this calculation deserve our attention. In this study, the carbonate sedimentation rate was assumed to be proportional to CO_2 partial pressure and more realistic dependences of surface air temperature on carbon dioxide content were used. It was found that the amount of CO_2 in the Precambrian atmosphere necessary to compensate for the influence of decreased solar radiation is less than the value obtained by Walker et al.

In this case the possibility of obtaining inadmissibly great changes in CO_2 content due to variations in the degassing rate was excluded.

Let us note that recent studies have treated the carbon dioxide and oxygen balance in external layers of the Earth, which is necessary to reveal the causes that restricted variations in atmospheric CO_2 and O_2 volume (Garrels et al. 1976; Holland 1978; Mackenzie and Pigott 1981 etc.). These studies show that initial data available up to now are insufficient to construct a detailed quantitative model describing atmospheric gas cycles that can be used to determine the chemical composition of the atmosphere in the geological past.

Nevertheless, there exists a real possibility of developing simplified models to solve this problem. In addition to the above-mentioned studies describing such models, we cite two recent studies which solve the problem of calculating variations in carbon dioxide content in the geological past. The first study (Berner et al. 1983) shows a method to reveal the relationship between carbonate sedimentation rate and carbon dioxide concentration similar to that used by Walker et al. The new idea of Berner et al. is that the rate of carbon dioxide production to the atmosphere is estimated by the rate of ocean floor spreading. Without touching other details of this study, we note that the corresponding dependency is probably ambiguous and the accuracy of data on spreading in different geological epochs for the ocean as a whole is rather limited. However, the authors of this paper could correctly conclude that over the last hundred million years, atmospheric CO_2 concentration tended to decrease, yielding a lower temperature.

Another method for geochemical calculation of atmospheric carbon dioxide concentration during the Cretaceous and the Tertiary periods has been developed by Lapenis (1984), who used data for different geological epochs on the level of carbonate compensation (i.e. the level at which the rates of carbonate sedimentation and dissolution are equal). Variations in this level are related to the balance of carbon compounds in the ocean, reflecting alterations in atmospheric carbon dioxide concentration (Byutner 1983).

Average CO_2 concentrations for the two epochs of the Cretaceous and the five epochs of the Tertiary obtained by Lapenis appeared to be similar to those calculated by the authors of this book and mentioned at the beginning of this section.

With several independent methods for determining variations in atmospheric carbon dioxide yielding similar results, we can conclude that these results are sufficiently reliable. However, this does not rule out the necessity of checking these results by valid empirical data rather than by calculations based on very simplified models. Among these empirical data are palaeotemperature estimates for the geological past which can give realistic information on variations in the amount of carbon dioxide in the atmosphere. The results of comparison of this information with calculated data are discussed in Chapter 3.1.

2 Methods for Determining Changes in the Composition of the Atmosphere

2.1 Sedimentary Layer of the Earth's Crust

Studies of Sedimentary Layer Changes in Time. In the concentric structure of the Earth the sedimentary shell (the stratisphere) forms the upper solid layer consisting of bedded sedimentary and volcanic rocks.

During the past decades, different authors have made attempts at estimating the total volume, mass and composition of the rocks making up the sedimentary shell. The accuracy of these estimates was very limited, since they were based either on general ideas or on only inadequate actual data. Practically nobody has considered quantitatively the history of formation and change of material in the sedimentary shell, which has restricted the possibilities of reconstructing the chemical composition of the atmosphere and the waters of the World Ocean in the past, since the traces of changes in their chemical composition could only remain in the stratispheric rocks that formed by interaction with the Earth's atmosphere and hydrosphere. These traces can be discovered in changes in mass and composition of the rocks that make up the sedimentary shell.

To interpret these traces, labour-consuming quantitative estimations of the parameters of the sedimentary layer and their changing tendencies in geological time were necessary. More than 30 years of systematic studies were required to solve this problem to a first approximation by summarizing empirical geochemical and geological information on the basis of quantitative estimates of volume, mass, abundance and chemical composition of the most important types of rock in the stratisphere and the principal features of their time variations. This became possible by the volumetrical investigation method (Ronov 1949), after estimation of its validity and limitations (Ronov et al. 1972, 1973), applied to the study of the Late Proterozoic and Phanerozoic lithological continental associations and oceans. This was done by measuring rock volumes, using the maps of world lithological associations of all geological systems since the Early Riphean (1600 million years ago) up to the present time (Ronov and Khain 1954; Ronov et al. 1984a; Ronov et al. 1983; Khain et al. 1983). Later many of these maps were revised in the light of new data, and by using their originals to scale 1:25,000,000, the volumes were measured repeatedly, some of the results being used in this book. At the same time the mass and chemical composition of the most important kinds of sedimentary and volcanic rocks were determined within the largest structural zones of the continents and the oceans at different stages of their development by using tens of thousands of rock chemical analyses by a standard technique in the Institute of Geochemistry and Analytical Chemistry (Academy of Sciences of the USSR) and partly taken from other studies (Amosov et al. 1980; 1983; Lisitsyn 1978; Ronov 1976; Ronov et al. 1965, 1966; Ronov and Migdisov 1970; Ronov et al. 1969; Ronov 1984b; Ronov et al. 1963; Ronov and Yaro- shevsky 1976; Trotsyuk 1979; Schwab 1973 etc.).

New quantitative geological and geochemical data have been used as the basis for many generalizations. They formed the basis for constructing a quan- titative model of the contemporary structure of the Earth's sedimentary shell, which allowed us to observe variations in the petrographical and chemical composition of erosion zones on the ancient continents, the evolution of the composition of sedimentation regions, the tendencies to time variations in the mobility of the Earth's crust and evolutionary transformations of the chemical composition of the atmosphere and oceanic waters determined by such varia- tions. One of the most important problems in Earth Science is to reconstruct these processes for the remote past of our planet by generalizing the extensive geological evidence available on a new quantitative basis.

Structure and Composition of the Sedimentary Layer. The Earth's sedimentary layer is not continuous. It wedges out to ancient shields and mid- oceanic ridges (Table 4). Of $149 \times 10^6 \text{km}^2$ of the continental and island area, $119 \times 10^6 \text{km}^2$, i.e. 80% of the total land area, are occupied by sedimentary shell.

Table 4. Characteristics of deposits of the largest structural zones of the continents, their margins, the oceans and the Earth's sedimentary shell as a whole: Area S and its ratio S' to the total Earth's surface area, volume V and its ratio V' to the total volume of sediments and average thickness

Global Structural Zones of the Earth	S 10^6km^2	S'%	V 10^6km^3	V'%	Average Thickness, km
Continental shields	30	5.9	0	0	0
Plates of continental platforms	65	12.7	227	20.6	3.5
Platforms as a whole	95	18.6	227	20.6	2.4
Geosynclines and oro- genic regions of the continents	54	10.6	523	47.6	9.7
Continents as a whole	149	29.2	750	68.2	5.0
Continental margins (shelves and conti- nental slopes)	64	12.6	250	22.7	3.9
Mid-oceanic ridges	71	13.9	0	0	0
Oceanic plates and abyssal basins	226	44.3	100	9.1	0.4
Ocean floor as a whole	297	58.2	100	9.1	0.3
Earth as a whole	510	100.0	1100	100.0	2.2

Crystalline rocks outcrop on the remaining part of the land. On the continents the sedimentary shell occurs on a "granitic" layer of the Earth's crust and is largely made up of Phanerozoic and Late Proterozoic rocks. More ancient, not greatly metamorphosed Proterozoic sediments of protoplatforms are known, however, they occur rather rarely over the continental surface. Sediments of the continental borders (shelves and continental slopes) are of similar age. They thin out to the oceans and occupy an area of $64 \times 10^6 km^2$.

The total oceanic pelagic area makes up $297 \times 10^6 km^2$, $226 \times 10^6 km^2$ being sediments of the first seismic layer occurring within oceanic plates and abyssal basins (76% of the total oceanic pelagic area) and the other $71 \times 10^6 km^2$ mid-oceanic ridges having practically no sedimentary cover. Here, on the floor surface, the rocks of the basaltic (the second seismic) layer are exposed.

Oceanic sediments lie on basalts of the second layer of the oceanic crust. They are considerably less in their stratigraphical coverage than the rocks of the continental sedimentary shell. The most ancient sediments discovered in deep-sea drilling by the *Glomar Challenger* (Initial Reports 1969-1981) refer to the Late Jurassic and are covered with deposits of the Cretaceous, Palaeogene, Neogene and Quaternary systems. Their age and thickness regularly decrease towards mid-oceanic ridges. As the age decreases, the distribution area increases.

So far it can only be guessed whether oceanic sediments more ancient than the Jurassic exist. Supporters of the plate tectonic theory believe that these deposits were accumulated in the geological past, but were later assimilated in subduction zones when the oceanic crust moved under the continents.

The thickness of the sedimentary shell varies widely from 0 to 20-30 or more kilometres, with maxima in geosynclinal zones, marginal platform basins of the Caspian type and in the shelf deep troughs. On the average for the entire Earth, the thickness of the sedimentary shell is only 2.2 km (Table 4). The average thickness in the global structure sequence decreases successively from the continents (5.0 km) to shelves and continental slopes (3.9 km) and then to the ocean floor (0.4 km). On the continents it decreases from geosynclines (9.7 km) to platforms (2.4 km).

The total volume of sedimentary shell is estimated as $1100 \times 10^6 km^3$, which makes up about 11% of the volume of the Earth's crust and 0.1% of the total volume of the Earth (Ronov and Yaroshevsky 1976). Such is the scale of the subject under consideration. The correctness of this estimate was later confirmed by calculations carried out using an independent technique by Southman and Hay (1981) and Khain et al. (1982). According to the results of these studies, the total volume of the stratispheric rocks is equal to $1115 \times 10^6 km^3$ and $1104 \times 10^6 km^3$. Agreement between these estimates and the previous one is surprisingly good. However, these authors present some discrepancy concerning the scheme of distribution of the total volume of rocks within the major global structures of the sedimentary shell.

In the total volume distribution of the sedimentary shell within the largest global structures of the Earth, there is, according to Table 4, a sharp disproportion, apparently reflecting the asymmetry and abyssal heterogeneity in the planet's structure (Fig. 1). A greater rock mass of sedimentary shell (68%) is

a)

Areas
$$S = 510 \times 10^6 \text{km}^2$$

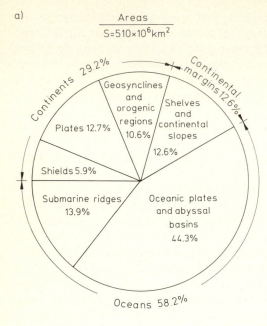

b)

Volumes
$$V = 1100 \times 10^6 \text{km}^3$$

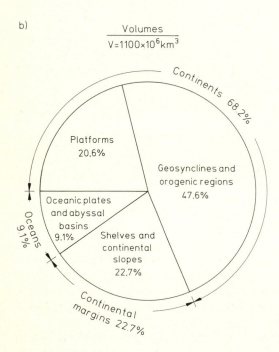

Fig. 1. The distribution of the Earth's total surface area (**a**) and the total volume of deposits in the Earth's sedimentary shell (**b**) among the continents, their margins and the oceans

contained on the continents, which occupy only 29% of the Earth's surface. The continental block, together with the shelf and continental slopes, includes 91% of the rock volume of sedimentary shell, covering in this case only 42% of the Earth's surface area. The floor of the oceans, occupying 58% of the total area of the planet, contains only 9% of the total volume of the stratisphere. This disproportion, with some variations, is retained in the schemes of other authors.

Irregularity in the distribution of volume and mass of the sedimentary shell occurs within every global structure. On the continents, for instance, there are two clearly defined systems of irregularity. The first is associated with the division of the continents into two blocks, Lavrasia and Gondwana, which differed from each other in the Neogäikum, i.e. in Phanerozoic and Late Proterozoic time, by a regime of vertical (epeirogenic) movements.

The second system of irregularities in the distribution of rock volume of the sedimentary shell of the continents is caused by the existence of tectonic zonation and differences in rates of sedimentation between platforms and geosynclines. More than two thirds of the total volume of the shell is concentrated in geosynclinal regions and only one third on the platforms (Table 4). The differences between stable and active zones of the continental crust are also evident in the proportions of their component rocks (Fig. 2). Volcanic and siliceous rocks are more widely represented in geosynclines, and carbonate rocks, salts, gypsum and anhydrites on the platforms. However, if we exclude volcanics, because they are not genetically associated with the sedimentary shell, these differences become greater (Table 5).

Significant irregularities in the distribution of volumes and associations have been observed through the vertical section of the sedimentary shell of the continents. They show up primarily in the absence of any relationship between the volumes of the rocks and the absolute durations of the largest Neogäikum stratigraphic units (Fig. 3). For example, almost half the total rock volume of the continental stratisphere (46%) is contained in the Palaeozoic sequences, which were deposited only during 335×10^6 years, i.e. for the interval of time making up 21% of the total duration of the Neogäikum (1.6×10^9 yr). In contrast, the Upper Proterozoic sequence occupies only 16% of the sedimentary shell volume but corresponds to a time interval well over half the total duration of the Neogäikum (64%). The heterogeneity revealed in sedimentary rock distribution depends on two factors: first, differences in the rates of sedimentation during different stages of the Neogäikum, and second, the processes of erosion and weathering, which could have removed part of the original rock volume. Investigations have shown that during the Phanerozoic the secondary processes of the destruction of rock masses on a global scale are not essential but their influence acquires great importance for the Late Proterozoic strata (Ronov 1980).

The proportions of the associations change significantly through the stratigraphic section of continental sedimentary shell. The changes are particularly evident, if we compare the Late Proterozoic and the Phanerozoic sections of the sedimentary shell (Fig. 4). In the Upper Proterozoic sequence, terrigenous rocks (79%) predominate, carbonate rocks (11%) and volcanic rocks (8%) have a limited development, and the amount of evaporites is insignificant (0.3%). In the

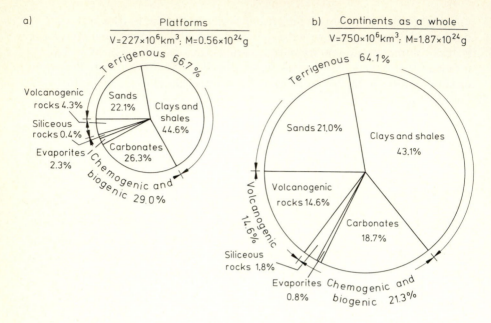

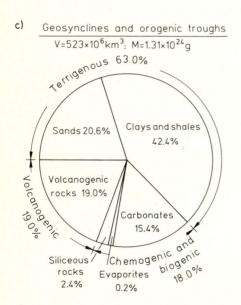

Fig. 2a-c. Volumes and occurrence of the most important rocks in the sedimentary cover of platforms **(a)**, geosynclines **(c)** and continental sedimentary shell as a whole **(b)**

Table 5. Volume, mass, average thickness and the ratio N of the volume of a given rock to the total volume of deposits for the most important types of Neogäikum rocks in the major structural zones of the Earth's sedimentary shell

Global structures	Structural zones	Sedimentary rocks with (and without) effusive rocks	Volume, 10^6 km^3	Mass, 10^{24} g	Average thickness, km	Sandy rocks	Clays and shales	Carbonate rocks	N, % Salts, gypsums, anhydrites	Siliceous rocks	Volcanogenic rocks
1	2	3	4	5	6	7	8	9	10	11	12
Continents	Platforms	Sedimentary and effusive rocks	227	0.56	2.4	22.1	44.6	26.3	2.3	0.4	4.3
		Sedimentary rocks only	217	0.53	2.3	23.1	46.6	27.5	2.4	0.4	–
	Geosynclines and orogenic depressions	Sedimentary and effusive rocks	523	1.31	9.7	20.6	42.4	15.4	0.2	2.4	19.0
		Sedimentary rocks only	423	1.03	7.8	25.5	52.4	19.0	0.2	2.9	–
	Continents as a whole	Sedimentary and effusive rocks	750	1.87	5.0	21.0	43.1	18.7	0.8	1.8	14.6
		Sedimentary rocks only	640	1.56	4.3	24.6	50.5	21.9	0.9	2.1	–
Shelves and continental slopes	Platforms	Sedimentary and effusive rocks	86	0.21	2.5	18.0	47.5	28.0	3.6	0.3	2.6
		Sedimentary rocks only	84	0.20	2.4	18.5	48.8	28.7	3.7	0.3	–

continued on page 40

Table 5 (continued)

Global structures	Structural zones	Sedimentary rocks with (and without) effusive rocks	Volume, 10^6km^3	Mass, 10^{24}g	Average thickness, km	Sandy rocks	Clays and shales	Carbonate rocks	N, % Salts, gypsums, anhydrites	Siliceous rocks	Volcanogenic rocks
1	2	3	4	5	6	7	8	9	10	11	12
Geosynclines and orogenic depressions		Sedimentary and effusive rocks	164	0.41	5.6	12.0	50.5	11.3	0.4	1.4	24.4
		Sedimentary rocks only	124	0.30	4.3	15.9	66.8	14.9	0.5	1.9	–
	Continental margins as a whole	Sedimentary and effusive rocks	250	0.62	3.9	14.0	49.4	17.2	1.5	1.0	16.9
		Sedimentary rocks only	208	0.50	3.2	16.9	59.5	20.6	1.8	1.2	–
Oceans		Sedimentary and effusive rocks	100	0.18	0.44	7.0	49.0	33.9	0.7	6.1	3.3
		Sediments only	96	0.17	0.42	7.3	50.7	35.0	0.7	6.3	–
Sedimentary shell of the Earth as a whole		Sedimentary and effusive rocks	1100	2.67	2.2	18.1	45.1	19.7	1.0	2.0	14.1
		Sedimentary rocks only	944	2.23	1.8	21.1	52.5	23.0	1.1	2.3	–

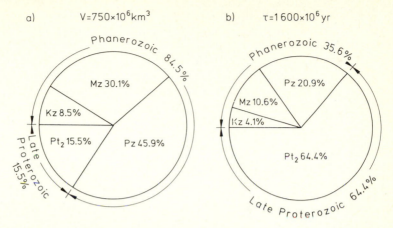

Fig. 3a,b. The distribution of rock volumes of continental sedimentary shell among large stratigraphic intervals of the Neogäikum **(a)** and their relative duration **(b)**

Phanerozoic sequences the proportion of terrigenous rocks is more moderate (62%), whereas the amounts of carbonates and volcanics are approximately twice as much as in the Proterozoic (20% and 16%, respectively) and the amount of evaporites is three times greater. The indicated features show that the process of differentiation of matter in the sedimentary shell became more intensive in the course of geological time.

Towards the continental margins and the oceans, the abundance of sedimentary rocks changes (Table 5): volumes of clay, carbonate and siliceous sediments increase and the volume of sandy rocks decreases.

The most important feature of sedimentary rocks is their clearly defined distinction in composition from the average composition of the rocks of the "granitic" shell, which have served as a major source of material for the sediments for at least the last 1.6×10^9 years. The difference shows primarily in the markedly higher content of water, CO_2 and organic carbon, as well as sulphur, chlorine, fluorine, bromine and other "excess" volatiles in the stratisphere and the directly associated hydrosphere. All geochemists agree that it indicates the direct release of those components from the mantle during degassing (Goldschmidt 1933; Rubey 1951; Vinogradov 1959, 1967).

Another important feature of sedimentary rock composition is its high calcium content[1] compared with the "granitic" layer, which has not yet been

---

[1]An increased calcium content in sedimentary rocks compared with the "granitic" shell has been attributed to different causes. The most probable is the following. On the basis of average values obtained in analyzing all the available granites, we judge the chemical composition of the "granitic" shell of the Earth's crust as revealed from geophysical data. At the same time, among the most ancient Archaean intrusive igneous rocks, anorthosites and endorbites, i.e. the rocks with calcium feldspar anorthite, prevail noticeably on all the continents. At the initial stages of existence of the Earth's sedimentary shell, these calcium (not potassium-sodium) intrusive igneous rocks were mainly eroded on the continents and islands, which gave great amounts of calcium to the stratisphere. At later epochs, besides igneous rocks, ancient limestone series were already eroded.

a)

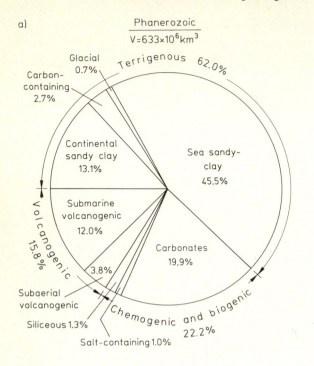

b)

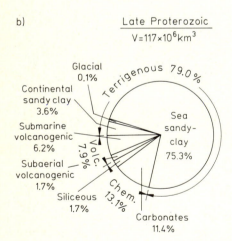

Fig. 4a,b. Comparison of the formation group volumes in the Phanerozoic **(a)** and Late Proterozoic **(b)** parts of continental sedimentary shell

explained in the geochemistry of the surface shells. Characteristic of sedimentary rocks is the sodium to potassium ratio shifted towards the potassium side, which is not compensated for by the excess of sodium in the ocean, which results in some deficit of sodium in the stratisphere and hydrosphere (taken together) relative to the "granitic" shell. In addition, sedimentary rocks are characterized by an increased ratio of ferric oxide to ferrous oxide, that is determined by oxidizing conditions on the Earth's surface, as well as an increased content of sulphate sulphur in sedimentary rocks (Ronov and Yaroshevsky 1976).

All these features are most clearly observed in the platform sediments, which are the products of deep weathering and intense surface differentiation. In contrast, the geosynclinal sediments have undergone substantially smaller changes (especially sands) and are closer in composition to the original parent rocks, among which basic volcanics of deep origin have played an extremely significant role. The role of the basic material becomes still more distinct if we consider that the contribution of the geosynclinal and particularly eugeosynclinal rocks to the overall balance of the sediments increases with increase in the geological age of the sedimentary sequences.

The validity of this has recently been confirmed by studies using extensive data on the modern continents which found that the most ancient clay rocks contain (compared to younger ones) a higher amount of elements typical of basaltoids (Mg, Cr, Ni, Co) and a lower content of elements peculiar to granitoids (K, Na, Rb). This is an indication that with time the contribution of major abyssal matter to the formation of sediments decreased and that of acid matter increased (Ronov 1984b). Only in abyssal sediments of the modern oceans do basic volcanism and weathering products of basic rocks play a leading role.

The average chemical composition of the sediments of layer I of the oceans differs noticeably from the average composition of the rocks of the sedimentary shell of the continents primarily by a higher content of carbonate components (CaO and CO_2), decreased concentrations of dispersed organic carbon (C_{org}) and such terrigenous components as Al_2O_3 and TiO_2, as well as by the ratio Na_2O/K_2O being displaced to the sodium side. The oceanic sediments are characterized by a predominance of montmorillonitic associations among the clay minerals, whereas in the sedimentary sequences of the continents hydromicaceous associations are predominant, especially in deep ancient horizons. Montmorillonite associations are common only in the upper levels of the section (Ronov 1980).

The Evolution of the Sedimentary Layer. Since the Archaean up to the present time, an irreversible process of growth of the platform areas and reduction in the geosynclinal ones has developed (Fig. 5). As a result of this, not only the area of distribution of geosynclinal volcanism and intrusive activity has decreased, but also a change in the intensity of weathering processes and removal of their products into the seas and the oceans has occurred. A comparative study of sedimentary differentiation within the stable and mobile zones of the crust established that the depth of disintegration of the rocks on the platforms is greater and the differentiation more complete. The value of removal decreases from the

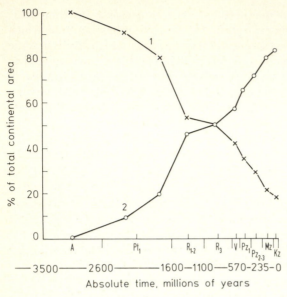

Fig. 5. A general trend to contracting geosynclinal area *(1)* and increasing platform area *(2)* within the modern continents

most mobile to the least mobile elements, forming a mobility series that corresponds to the distribution of these elements in the waters of the present ocean:

Na > Mg > Ca > K.

During weathering of geosynclinal rocks, this series is modified as a consequence of the more basic composition of the parent rocks and the incompleteness of their disintegration in weathering:

Mg > Ca > Na > K.

It can be assumed that at the early stages of the crust's development, when the geosyncline area was much greater than the platform one, the processes of removal of elements into the ocean followed the geosynclinal model, whereas at later stages they followed the platform model. On the basis of these conclusions and observation of modern natural and experimental models of weathering (hydrolysis) of granites and basalts, a rough scheme of the evolution of the cation composition of the World Ocean water has been outlined. It follows from this scheme that waters of the ancient Early Precambrian ocean differed from modern sea water by a greater content of Ca, Mg, K and a lesser content of Na.

The age of the most ancient metamorphic sedimentary rocks approaches 3.8×10^9 years. It is probable that still older sediments exist, traces of which may be discovered in the lower horizons of "granitic" and in "basaltic" crust layers. This assumption is confirmed by recent studies of Australian geologists, who found redeposited zircons of an age approaching 4.2×10^9 years in the Early Precambrian metamorphic strata of the West Australian shield.

The position of the upper and lower boundaries of the sedimentary shell has changed during geological time. A rise of the lower boundary of the shell was associated with the processes of regional metamorphism and granitization of the sedimentary sequences, which developed in the deep zones of the mobile belts, finally leading to accretion of the "granitic" shell on the continents. The region in which these processes occur has decreased in area with each passing cycle since the end of Archaean time as a consequence of the stabilization of the crust and the growth of the platform areas on the sites of the geosynclines that terminated their development (Fig. 5). The process of reduction of the area of geosynclinal regions seemed to be dependent on the total decrease in the amount of radiogenic heat that has been released since the initial stages of our planet's development up to the present epoch.

Rough estimates show that the total contribution of sedimentary material to the Earth's crust (including metamorphic sediments of the Precambrian) comprises not less than 35% of its total mass.

The position of the upper boundary of the sedimentary shell has also changed over time as a consequence of the deposition of younger and younger series of sediments in the regions of crustal subsidence. The growth of the shell resulted from the influx of volcanogenic material from the Earth's depth, terrigenous material from the shields and inner rises and chemogenic material from the oceans and the atmosphere; all resources were continuously replenished by the removal of dissolved products of weathering from the continents and of "excess" volatiles from the Earth's interior.

At the present epoch, the World Ocean (including continental and marginal seas) covers $361 \times 10^6 \text{km}^2$, i.e. about 71% of the Earth's surface. This is its minimal area, since we live in a geocratic epoch characterized by sea regression from elevated continents. Geocratic epochs have also occurred in the past and were associated with the final stages of geotectonic cycles (Fig. 6). Among them are the Early Devonian — the boundary of the Caledonian and the Hercynian cycles, Late Permian-Middle Triassic — the boundary of the Hercynian and the Alpine cycles, the Miocene-Quaternary period — the boundary of the Alpine and the future cycle which has as yet no name. The areas of intracontinental seas were always minimal in geocratic epochs. During thalassocratic epochs, associated with the middle stages of cycles, the areas of intracontinental seas considerably increased, mainly due to continental subsidence and the development of large transgressions. These thalassocratic cycles were the Ordovician, the Middle Devonian-Early Carboniferous and the Late Cretaceous. At the same time, there was a similar periodicity in the development of large transgressions and regressions on the platforms and geosynclines. Differences between these zones were only quantitative. During a greater part of the Phanerozoic, the seas covered more than half the total area of geosynclines and less than 25% of the area of platforms.

The theory of geocratic and thalassocratic epochs is based on statistical data on all continental areas where sea sediments of differing age are developed. This

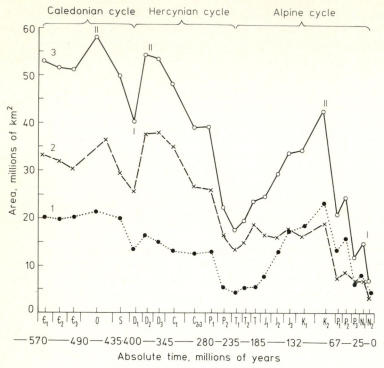

Fig. 6. Time changes in area covered with sea within the platforms *(1)*, geosynclines *(2)* and the modern continents as a whole *(3)* Epochs: *I* geocratic; *II* thalassocratic

does not mean that transgressions occurred simultaneously everywhere. In particular, a vast transgression on the Australian continent is recorded in the Early Cretaceous and not in the Late Cretaceous.

Periodic uplift and subsidence of the surface of the continents entailing transgression and regression of the adjoining sea and climatic changes induced a varied scale of weathering intensity and erosion on the continents, volume of fluviatile supply, and mass and proportion of mechanical and dissolved products transported from the continents to the oceans. For this reason, great care is necessary when using the parameters observed in the present epoch to calculate geochemical mass balances ("fluxes") during the geological past. At best, the present time may serve as a standard for the geocratic periods of the Phanerozoic. It is extremely doubtful whether current parameters are a completely reliable measure for the epochs with major transgressions, not to mention the geological periods of the remote past, when the petrological composition of land area differed greatly from that at present.

These doubts can be confirmed by quantitative analysis of the ratio of the most important groups of lithological associations of the continents for different epochs and periods of the Phanerozoic. This ratio clearly expresses periodic changes in their proportions in the course of geotectonic cycles. The maximum deposition of terrigenous clastic associations, in particular, those of continental origin, occurs during the concluding (orogenic) phases of the Caledonian, Hercynian and Alpine cycles as a result of high elevation of land areas, marine regression, and predominance of uplifts over continental subsidence. Maxima in the amount of carbonate formation occur during the middle phases of cycles, in the epochs of great transgression and relatively slow uplift of eroded regions and subsidence of sedimentation regions of the continents. Changes in total volume of continental sediments with maxima in the middle of geotectonic cycles and minima in their initial and terminating phases are also cyclic in nature. The results of measuring rock volumes indicate not only cyclic development of the process of subsidence of the Earth's continental crust but also to the conformity in movements of geosynclines and platforms, as well as to much greater scales of sedimentation and subsidence in mobile zones of the Earth's crust than in the stable ones (Fig. 7).

Volumes of sediment on the platforms and in the geosynclines change in the same direction in time. This indicates their origin from global processes which cause periodic rejuvenation and attenuation of tectonic activity. This is generally reflected in transgression and regression of the World Ocean, increase or decrease of sediment volume and of average rates of subsidence.

This global rhythm exists in spite of the differences in behaviour of individual major lithospheric blocks, in particular the continental platforms (Yanshin 1973),

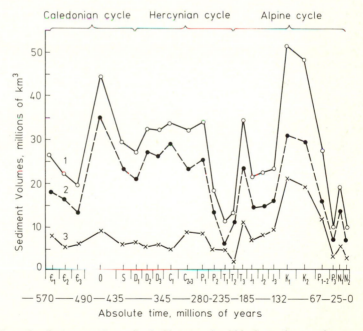

Fig. 7. Time changes in total sedimentary volumes for the Phanerozoic within the modern continents *(1)*, continental platforms *(3)* and geosynclines *(2)*

which indicates the predominant role of global rather than regional trends. At the Phanerozoic stage, irreversible and directed changes in palaeogeographic conditions of sedimentation appeared, which resulted in an increase in the area of continental sedimentation and in a corresponding decrease of the marine sedimentation area (Fig. 8).

Let us now consider the evolution of the petrological composition of continental erosion regions (Fig. 9). The major trend in this evolution was a successive reduction in the area of outcrop of basic volcanics and an increase in the area of sedimentary rocks. The distribution of granitoids is more complex. With the intensification of granitization and subsequent erosion of ancient fold regions, the area of their outcrop increased, reaching maximum before the beginning of the Late Proterozoic. Then it gradually decreased as the crystalline basement was covered with Late Proterozoic and younger sediments.

Proceeding from the accepted scheme of time changes in the proportions of eroded rocks of the continents and using data on the composition of rocks of different ages from the USSR, USA and other countries, geochronological trends were considered in variations of the average chemical composition of the eroded substrata of the continents. The geochemical consequence of transformations of the ancient land composition was a change in the composition of terrigenous products of weathering and the solutions transported from continental erosion regions into intracontinental seas and oceans. This definitely affected the overall direction of the evolution of sediment composition and of the chemical composition of oceanic waters.

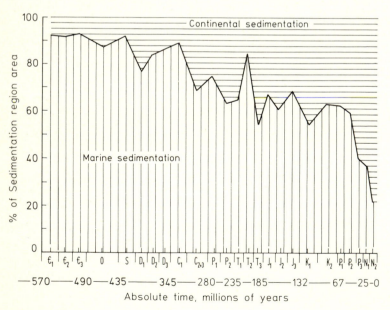

Fig. 8. Change in the ratio between sea and continental sedimentation areas within the modern continents during the Phanerozoic

Fig. 9. Time changes in ratios of the most important rock groups in the continental erosion regions. *1* granitoids and orthogneis; *2* lava (mainly basic), *3* sedimentary rocks

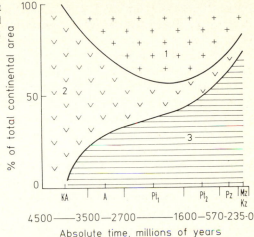

Absolute time, millions of years

Let us now treat some essential details of the evolution of lithological composition and proportions of sedimentary and volcanic rocks in the regions of sedimentation on the continents (Fig. 10). Firstly, we shall consider carbonate rocks .These are comparatively rare in the Archaean. During the Early Proterozoic their abundance markedly increases, reaching maximum in the Late Proterozoic and Palaeozoic. At this time dolomites prevail among the carbonates, whereas limestones prevail in the Mesozoic-Cenozoic interval. It is to be noted that in the Early Precambrian evaporites are almost absent. Thus, calcium sulphates (gypsum and anhydrites) seem to appear only in the second half of the

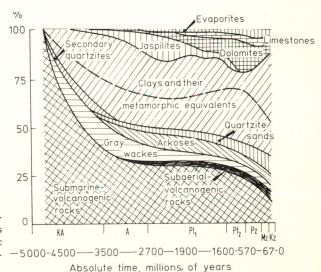

Fig. 10. The evolution of lithological composition and ratios between sedimentary and volcanic rocks in the continental sedimentation regions

Absolute time, millions. of years

Early Proterozoic, and rock salt in the Vendian, i.e. at the end of the Late Proterozoic. Evaporites were precipitated most abundantly during Palaeozoic time.

Within each of these rock types irreversible qualitative changes in lithological and mineral composition occurred (Yanshin et al. 1977). These changes have been studied in detail in the evaporite associations by Zharkov (1974, 1978) and in the red bed associations by Anatoljyeva (1972).

The trends in changes with time of the average chemical composition of sedimentary rocks of the stratisphere (in particular, the argillaceous types) revealed after summarizing data of numerous analyses of basement rocks and sedimentary covers of the Russian, North American, Siberian and other ancient platforms of the world, show that the distribution of potassium and sodium in the sediments is essentially the same as that in rocks in the erosion zones on the continents. The sodium content of sedimentary rocks decreases from Archaean to Mesozoic time. This trend is due to interaction of two genetically different processes identically directed in time. The first process was controlled by the closure of the Precambrian geosynclines and the expansion of the platform area. A sequential decrease in effusive volcanism led to a gradual reduction in the influx of deep-seated material relatively enriched in sodium to the platformal regions of sedimentation. The second process was associated with the cyclic weathering and sedimentation of the original Precambrian geosynclinal material. As a result, the terrigenous material was gradually depleted in soluble sodium, which was carried away from the continents by the river systems and was accumulated in the oceans. The potassium content increased until the Early Palaeozoic, then decreased markedly in the younger sediments. This trend is a function of the interrelated processes of granitization, subsequent erosion of the granitized rocks, deposition of arkosic association in the platform sands and incorporation of potassium by clay minerals (mainly hydromicas). As the basement and a significant part of the shields were covered with Palaeozoic sediments, the transfer on the platforms of terrigenous material enriched in potassium started to decrease. The passage of the sediments through repeated cycles of weathering and sedimentation led at the end of the Palaeozoic era and during Mesozoic and Cenozoic times to the removal of potassium from sands and clays as a result of the breakdown of potassium feldspars in the arenaceous rocks and the replacement of hydromicas in clays by mont-morillonite.

The data presented here show convincingly that the role of ancient sedimentary rocks as a source of material for younger sediments increased with time, while that of volcanics and granitoids, vice versa gradually decreased during the Phanerozoic. A decrease in the contribution of basic material during sedimentation was reflected in the process of removing iron from sedimentary rocks. Accumulation of ferruginous rocks (jaspilites) during the Early Protero-zoic interval was preceded by a long period of exogenic reworking of the subsequently granitized Archaean volcanic-sedimentary rocks, which initially contained a large amount of iron. This was reflected not only in ore formation, but also in an increased level of the average amount of iron in the synchronous Lower

Proterozoic meta-argillaceous sequences of the Baltic, Ukrainian, Aldanian, Canadian and Brasilian shields. Subsequently the composition of argillaceous rocks was accompanied by depletion in iron. The iron content of clays of the Russian, Siberian and North American platforms was almost halved during the period from the Early Proterozoic to the Mesozoic and the Cenozoic.

The depletion of sedimentary rocks in iron was accompanied by the separation of iron and manganese. This can be explained only if iron, which is more readily oxidized than manganese, turned into a less mobile trivalent form with an increasing amount of free oxygen in the atmosphere resulting in the separation of these elements. The same process was invoked to explain the origin of the greatest manganese deposits on the continents, formed at the end of Phanerozoic time (in the Palaeogene), which are almost free of iron (Nickopol and Chiatury) and iron deposits poor in manganese (the Lorraine basin and the Kerchian peninsula).

In the 1950's, a permanent trend was found which showed an increase in the Ca/Mg ratio with time in carbonate rocks (Vinogradov and Ronov 1956). In the light of new data, this trend proved to be characteristic of not only carbonates but also, in general, of all continental sediments. It is a function of changes in the composition of regions of continental supply; this is proved by a simultaneous increase of the Ca/Mg ratio in parent rocks and in combined sediments (Ronov 1972).

Among other irreversible trends in changes in the composition of sedimentary rocks we should mention an overall increase in the amount of organic carbon, sulphate and pyritic sulphur and in the ferric oxide to ferrous oxide ratio. As the mass of organic matter disseminated in sediments increased, the iron of their minerals was not reduced but actually increasingly oxidized. This is confirmed by the distribution of sulphate and pyritic sulphur. Proceeding from these facts, and taking into account that the bulk of clay minerals formed in weathering zones in contact with the atmosphere, we properly conclude that the intensity of oxidizing processes increased with geological time at the Earth's surface and the atmospheric content of free oxygen also increased. At the same time, the atmospheric CO_2 concentration gradually decreased, the nitrogen content increased, and the amount of oxygen produced mainly in photosynthesis also grew. These trends in the evolution of atmospheric gas composition predetermined in many respects the manner of development of the anion composition of ancient oceanic waters. The growth of oxygen concentration in the atmosphere affected primarily the history of sulphate ion. Oxidation of sulphur to sulphate resulted at the end of the Proterozoic to the beginning of the Palaeozoic in a considerable increase in the sulphate concentration of sea water, and changed the type of equilibria that controlled the content of sulphur in oceanic water. Decreasing atmospheric CO_2 partial pressure was accompanied by simultaneous reduction of the bicarbonate and carbonate ion concentration in oceanic waters. At the same time, the amount of chlorine in oceanic waters increased gradually as it was carried out from the depths of the Earth into its upper shells. Accumulation of chlorine was promoted by its long average residence time in the ocean.

Thus, major trends in changes of atmospheric gas composition and ocean water composition are interrelated with the evolution of the sedimentary shell of the Earth, with life development and degassing of the Earth's depth. The sedimentary shell, ocean and the atmosphere represented a complicated system, throughout the entire geological history, whose time changes were determined by the dynamics of matter exchange between the components of individual systems and between the entire system and the Earth's depths.

2.2 Carbon in the Sedimentary Layer

The Balance of Organic and Carbonate Carbon. Simultaneous analysis of new and earlier data allowed us to consider to a first approximation the balance of carbonate and organic carbon in the sedimentary shell of the Earth as a whole. The Neogäikum sediments of the continents have been analyzed. This interval of time covers several geological eras of 1.6×10^9 years in total duration. A time period ten times shorter, equal to 150 million years, has been considered for the oceans, since the deep-sea drilling revealed here only Late Mesozoic and Cenozoic sediments. The supporters of the theory of new global tectonics believe that more ancient oceanic sediments were destroyed as a result of their subduction under active continental margins. Considering this, we will cite the results of calculating the terms of the C_{carb} and C_{org} balances in the following two variants: without and with taking into account the effects of subduction processes on the estimation of C_{carb} and C_{org} volumes in pelagic oceanic deposits.

The first variant of carbon balance is presented in Fig. 11. The major reservoirs are depicted as squares, the sizes of which are proportional to the carbon mass in each of them. Figure 11a represents the C_{carb} balance model. Its total mass in the stratisphere, including not only limestone and dolomite series but also carbonate admixture in other rocks, comprises 860×10^{20} g. As can be seen in Fig. 11, the distribution of C_{carb} in the three reservoirs is unequal in volume. The largest is represented by continental sedimentary rocks that contain 563×10^{20} g of C_{carb}, or 66% of its total mass in the sedimentary shell. The reservoir of shelf and continental slope deposits is 2.5 times less and contains 224×10^{20} g of C_{carb}, or 26% of its total amount in the shell. Oceanic pelagic sediments containing 73×10^{20} g of C_{carb}, or 8% of its total mass in the stratisphere are the smallest reservoir.

Figure 11b shows the balance of C_{org} dispersed in the sedimentary shell. Its total volume is equal to 118×10^{20} g, seven times less than that of C_{carb}. The continental sediments are the most capacious reservoir for dispersed organic carbon as well as for C_{carb}, and contain 83×10^{20} g of C_{org}, or 71% of its total volume in the sedimentary shell. This estimate is based on empirical data on the total content of C_{org} in the whole stratigraphic complex of the Neogäikum. It agrees well with the estimate derived from data on an average content of C_{org} in the stratisphere (0.52%). The reservoir of C_{org} in shelf and continental slope sediments is 2.5 times less: it makes up 33×10^{20} g, or in other words 28% of

a)

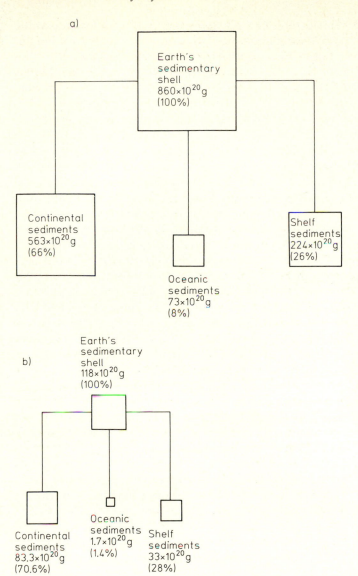

Earth's
sedimentary
shell
860×10^{20}g
(100%)

Continental
sediments
563×10^{20}g
(66%)

Oceanic
sediments
73×10^{20}g
(8%)

Shelf
sediments
224×10^{20}g
(26%)

b)

Earth's
sedimentary
shell
118×10^{20}g
(100%)

Continental
sediments
83.3×10^{20}g
(70.6%)

Oceanic
sediments
1.7×10^{20}g
(1.4%)

Shelf
sediments
33×10^{20}g
(28%)

Fig. 11a,b. The C_{carb} **(a)** and C_{org} **(b)** balances in the sedimentary layer of the Earth's crust

the total mass of organic carbon in the sedimentary shell. The volume of the reservoir of oceanic pelagic sediments is negligibly small: it equals 1.7×10^{20}g of C_{org}, which comprises only 1.4% of its total mass in the stratisphere.

The ratio between C_{org} and C_{carb} masses amounts to 1:7.3 for the Earth's sedimentary shell, 1:6.8 for the continents, 1:43 for the oceans and 1:6.8 for the shelves and continental slopes.

If we hold to the idea of mobility and assume that in the past geological periods subduction took place, then we will come to the conclusion that the amount of pelagic sediments absorbed in the zones of Zavaritsky-Beniof as oceanic crust moved under the continents should increase on increasing the age of these sediments. The results by Lisitsyn (1980) showed that over the last 150 million years, i.e. since the Late Jurassic up to the end of the Quaternary period, 225×10^{6}km^3 of pelagic sedimentary material moved to the subduction zones. Agreeing with this estimate and adding to it the volume of preserved pelagic sediments of the same age (100×10^{6}km^3), we assume that the total volume of pelagic sediments (preserved or destroyed) for the time period from the Late Jurassic to the Quaternary period should have reached 325×10^{6}km^3 and their volume 549×10^{21}g.

Now we shall try to determine to a first approximation the C_{carb} and C_{org} masses in total preserved and destroyed pelagic sediments of the oceans. Let us assume in this case that the ratio between the volume of carbonate and other pelagic sediments in their destroyed portion was the same as in the preserved one, assuming as well an average content of C_{org} in sediments.

According to our calculations, the mass of carbonates and carbonate admixtures in the preserved pelagic deposits is equal to 97×10^{21}g, which is equivalent to 55% of their total mass (177×10^{21}g). Based on this value (55%), we find the mass of carbonate matter in the preserved and destroyed oceanic pelagic sediments to be in total 302×10^{21}g, and the mass of CO_2 and C_{carb} 133×10^{21}g and 360×10^{20}g, respectively.

The average relative content of carbonate matter in pelagic deposits was somewhat smaller (49%) according to Lisitsyn's data. Comparison of the estimates obtained shows that our results and those based on Lisitsyn's data are almost identical.

Similarly, we have calculated the mass of dispersed organic carbon in the total preserved and destroyed pelagic oceanic sediments. An average content of C_{org} in the preserved sediments comprises 0.11% and its mass 1.73×10^{20}g. Substituting these values into the total mass of the preserved and destroyed oceanic pelagic sediments, we find that since the Late Jurassic up to the end of the Pliocene, 6.0×10^{20}g of C_{org} have been buried in the sedimentary shell of the ocean.

As has already been mentioned, the ratio between the masses of C_{org} and C_{carb} is not constant. Its value is minimum in the reservoir of pelagic sediments of the oceans; however, for the continental block reservoirs, i.e. for the continents and their margins, it is six times greater and equals 1:7.

The ratio between the C_{org} and C_{carb} masses also changes with time. Figure 12 shows a distinct regular tendency towards an increase in the relative role of organic carbon from the Late Proterozoic to the Cenozoic deposits. Its direction in time has been determined by the interaction of two variable factors: first, a sequential increase in the average amount of organic carbon in the sedimentary rocks (from older to younger), and second, a decrease in the amount of carbonate carbon after the maximum in the Middle and Late Palaeozoic (Fig. 13).

In the Earth's history the release of "excess volatiles" such as water, sulphur, chlorine, fluorine, boron and CO_2 from the depths to the external layers of the planet took place. The mechanism of controlling the scales of carbonate accumulation, the resources of the biomass of organisms and buried organic matter at the Phanerozoic stage have earlier been considered quantitatively (Ronov 1982

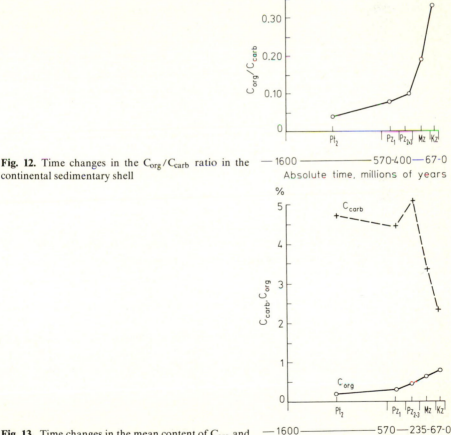

Fig. 12. Time changes in the C_{org}/C_{carb} ratio in the continental sedimentary shell

Fig. 13. Time changes in the mean content of C_{org} and C_{carb} in the continental sedimentary shell

etc.). It was shown that the measure of the intensity of volatile component release (including CO_2) from molten volcanic masses in one or another epoch is to a first approximation the volume of submarine and subaerial volcanic eruptions. Comparison between the volume of carbonate and volcanogenic rocks and the area of continental seas obtained from measurements by maps of the world lithological associations showed that during the entire Phanerozoic a direct relationship existed between the volumes of submarine and subaerial effusions, on the one hand, and the volume of accumulated carbonate rocks, on the other. The similarity of the curves describing changes in the volume of volcanogenic rocks and CO_2 of carbonates (Fig. 14) shows that as volcanogenic carbon dioxide comes into the atmosphere and the ocean, this gas was simultaneously, from the geological point of view, extracted from the indicated reservoirs and deposited in the form of carbonate sediments mainly to the floor of shallow platform and geosynclinal continental seas. This figure also demonstrates that a relationship existed between the area covered with sea and the volume of accumulated carbonate rocks. As the continental and shelf seas expanded, the amount of carbonates increased; and, vice versa, a decrease in their area caused a proportional decline in the volume of synchronous carbonates. All this led to the

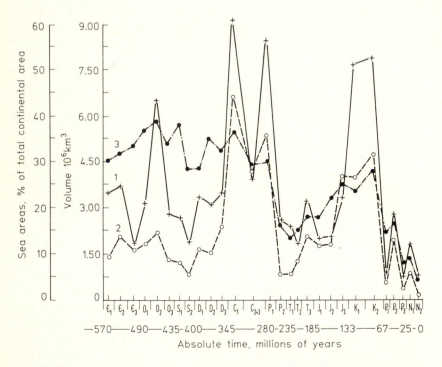

Fig. 14. Time changes in volcanogenic rock volumes *(1)*, CO_2 buried in synchronous carbonate rocks *(2)* and the ratio (%) of the continental sea area to the total area of the continents *(3)*

conclusion that changes in the carbonate rock volume were determined by corresponding periodic changes in the intensity of accordant volcanic and tectonic processes. The former determined the quantity of CO_2 necessary for carbonate deposition and the latter determined the area of the continental seas favourable for the deposition of carbonate sediments. In this connection it has been assumed that the quantity of carbonate sediments deposited during any given epoch after the Precambrian was directly proportional to the intensity of volcanic activity and the area of the distribution of the continental seas (Ronov 1959).

Our new data show that the fundamental role of carbonate deposition is a particular example of the more general rule of depositing carbon in the Earth's sedimentary shell (Fig. 15). The curve for the distribution of masses of residual organic carbon follows the general trend of the curves that reflect the changes in the mass of the volcanic rocks and the total mass of CO_2 in carbonate rocks and in the carbonate portions of non-carbonate rocks. This means that both organic carbon and carbonate carbon were deposited at the expense of the same source,

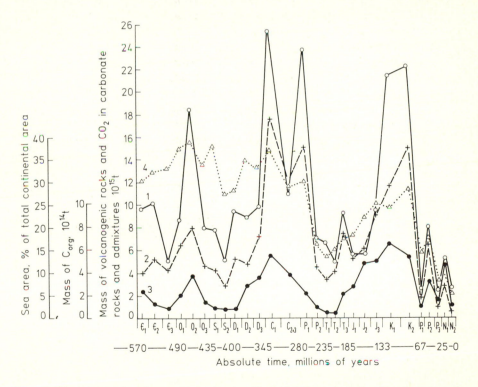

Fig. 15. Time changes in volcanogenic rock mass *(1)*, total CO_2 content of carbonate rocks and carbonates contained in other rocks *(2)*, C_{org} buried in the continental sedimentary strata *(3)* and the ratio (%) of the continental sea area to the total area of the continents *(4)*

and were subjected to the general rule of distribution. The carbonate rocks and the residual organic matter are consequently two derivatives of a single process of transport of deep-seated carbon dioxide to the surface.

Calculations have shown that with a limited mobile reserve of carbon in the equilibrium system: atmosphere-ocean-biosphere, carbonate deposition and organic carbon deposition should, from the geological point of view, have exhausted the carbon dioxide resources instantaneously, if the mechanism of its transport from the Earth's depths had not operated. Figure 15 not only confirms the existence of this mechanism in the geological past, but also shows that its intensity has varied through time.

The corresponding periodic changes in the intensity of the processes are clearly reflected in the form of peaks and minima on the appropriate curves. The epochs of stress volcanism and transport of vast masses of deep carbon dioxide, which are confined to the middle phases of the tectonic cycles (O, D_3-C_1, K), correspond to intense carbonate deposition and maximum fixation of residual organic matter in the sedimentary sequences. Meanwhile, the epochs of marked weakening of volcanic activity and reduced transport of CO_2 to the surface, which concentrate towards the initial and concluding phases of the cycles ($\varepsilon,S-D$, P_2T_2, $P-N_2$), correspond to the manifold decrease in the mass of deposited carbonate sediments and buried organic matter.

The alternation of epochs of enrichment and depletion in the sedimentary sequences in residual organic matter seems to reflect fluctuations of the total mass of the organisms that inhabited the Earth during any particular epoch. These fluctuations have been controlled by two global factors: changes in physicogeographical conditions of the biotic environment (the area of the seas, the relief of the Earth's surface and the climate) and changes in the intensity of volcanism, which determined the mass of deep carbon dioxide transported to the surface. The periodic changes in the conditions of the physicogeographical environment on the Earth's surface have been determined, directly or indirectly, by the overall course of the evolution of crustal movements during a tectonic cycle. A direct relationship exists between the rate of subsidence of a geosyncline and the intensity of volcanic activity in geosynclines: as the rate of subsidence increases, so the intensity of volcanism increases, and conversely, as the rate decreases, so the intensity also diminishes. Consequently, effusive volcanism of geosynclines has been clearly associated with the regime of epeirogenic movements.

In conclusion, we present tables characterizing the masses of carbonate carbon and organic carbon for different time intervals of the Phanerozoic. Table 6 includes data pertaining to the continents and Table 7 to the sedimentary shell as a whole.

The Balance of Oxygen. Let us now direct our attention to another problem connected with the accumulation of organic carbon in the rocks of the stratisphere, i.e. to the problem of oxygen balance in the external shells of the Earth. As is known, Vernadsky pointed out (1934) that the output of organic carbon from

Table 6. Volume of components of the continental sedimentary shell

Stratigraphic interval	Duration of intervals (m.y.)	Area of seas in % of total area of continents	Mass, 10^{21} g						Mass per unit of time 10^{20} g/10^6 yr	
			Volcanogenic rocks	CO_2 of carbonate rocks	CO_2 of carbonate admixtures to other rocks	Total mass of CO_2	C_{carb}	C_{org}	C_{carb}	C_{org}
1	2	3	4	5	6	7	8	9	10	11
L.Cambrian	570–545=25	30	9.72	3.53	0.43	3.96	1.08	0.23	0.43	0.09
M.Cambrian	545–520=25	32	10.25	5.01	0.28	5.29	1.44	0.14	0.58	0.06
U.Cambrian	520–490=30	33	5.04	4.03	0.10	4.13	1.13	0.08	0.38	0.03
L.Ordovician	490–475=15	37	8.68	4.54	1.91	6.45	1.76	0.21	1.17	0.14
M.Ordovician	475–450=25	39	18.34	5.54	2.32	7.86	2.14	0.38	0.86	0.15
U.Ordovician	450–435=15	34	7.78	3.28	1.38	4.66	1.27	0.17	0.85	0.11
L.Silurian	435–415=20	38	7.67	3.15	1.14	4.29	1.17	0.10	0.58	0.05
U.Silurian	415–402=13	28	5.07	1.99	0.73	2.72	0.74	0.08	0.57	0.06
L.Devonian	402–378=24	28	9.35	4.03	1.08	5.11	1.39	0.08	0.58	0.03
M.Devonian	378–362=16	35	8.54	3.78	1.01	4.79	1.31	0.29	0.82	0.18
U.Devonian	362–346=16	33	9.58	6.05	1.00	7.05	1.92	0.36	1.20	0.22
L.Carboniferous	346–322=24	37	25.62	16.78	0.72	17.50	4.78	0.55	1.99	0.23
M. and U. Carboniferous	322–282=40	29	10.95	10.71	1.18	11.89	3.24	0.38	0.81	0.09
L.Permian	282–257=25	30	23.74	13.38	1.75	15.13	4.13	0.22	1.65	0.09
U.Permian	257–236=21	16	6.97	2.19	2.33	4.52	1.23	0.10	0.59	0.05
L.Triassic	236–221=15	13	6.62	2.20	1.11	3.31	0.90	0.07	0.60	0.05

continued on page 60

Table 6 (continued)

Stratigraphic interval	Duration of intervals (m.y.)	Area of seas in % of total area of continents	Mass, 10^{21} g						Mass per unit of time 10^{20} g/10^6 yr	
			Volcanogenic rocks	CO_2 of carbonate rocks	CO_2 of carbonate admixtures to other rocks	Total mass of CO_2	C_{carb}	C_{org}	C_{carb}	C_{org}
1	2	3	4	5	6	7	8	9	10	11
M.Triassic	221–211 = 10	15	4.96	3.20	0.88	4.08	1.11	0.04	1.11	0.04
U.Triassic	211–186 = 25	18	9.01	5.17	1.74	6.91	1.88	0.22	0.75	0.09
L.Jurassic	186–168 = 18	18	5.57	4.79	0.54	5.33	1.45	0.28	0.80	0.15
M.Jurassic	168–153 = 15	22	5.71	4.91	0.88	5.79	1.58	0.49	1.05	0.33
U.Jurassic	153–133 = 20	26	9.27	10.19	1.12	11.31	3.08	0.49	1.54	0.25
L.Cretaceous	133–101 = 32	24	21.39	9.83	1.77	11.60	3.17	0.66	0.99	0.21
U.Cretaceous	101–67 = 34	28	22.06	11.87	3.02	14.89	4.06	0.54	1.19	0.16
Palaeocene	67–58 = 9	14	2.46	1.46	0.22	1.68	0.46	0.08	0.51	0.09
Eocene	58–37 = 21	17	7.78	4.78	1.45	6.23	1.69	0.31	0.80	0.15
Oligocene	37–25 = 12	8	2.18	0.69	0.24	0.93	0.25	0.15	0.21	0.13
Miocene	25–9 = 16	9	5.05	2.22	0.80	3.02	0.82	0.31	0.51	0.19
Pliocene	9–2 = 7	5	2.38	0.45	0.32	0.77	0.21	0.09	0.30	0.13
	570–2 = 568	271.74	149.75	31.45	181.20		49.39	7.10	0.87	0.13

Table 7. Volume of components of the sedimentary shell as a whole

Stratigraphic interval	Duration of intervals (million years)	Global structure	Mass, 10^{21} g				Mass per a unit of time 10^{20}g/10^6 yr	
			Total mass of sedimentary rocks	Volcanogenic rocks	C_{carb}	C_{org}	C_{carb}	C_{org}
1	2	3	4	5	6	7	8	9
Jurassic	153−133 = 20	Continents	47.03	9.27	3.08	0.49	1.54	0.25
		Shelves and continental slopes	33.76	12.07	0.88	0.30	0.44	0.15
		Floors of oceans	2.49	0	0.20	0	0.10	0
		Sedimentary shell as a whole	83.28	21.34	4.16	0.79	2.08	0.40
Cretaceous	133−101 = 32	Continents	107.75	21.39	3.17	0.66	0.99	0.21
		Shelves and continental slopes	76.65	28.14	1.72	1.01	0.54	0.32
		Floors of oceans	18.36	0.66	0.47	0.04	0.15	0.01
		Sedimentary shell as a whole	202.76	50.19	5.36	1.71	1.68	0.54
Cretaceous	101−67 = 34	Continents	98.28	22.06	4.06	0.54	1.19	0.16
		Shelves and continental slopes	63.58	23.39	1.55	0.18	0.46	0.05
		Floors of oceans	30.04	1.19	1.19	0.03	0.35	0.01
		Sedimentary shell as a whole	191.90	46.64	6.80	0.75	2.00	0.22
Palaeocene	67−58 = 9	Continents	14.13	2.46	0.46	0.08	0.51	0.09
		Shelves and continental slopes	12.60	3.26	0.37	0.02	0.41	0.02
		Floors of oceans	7.16	0.56	0.42	0.01	0.47	0.01
		Sedimentary shell as a whole	33.89	6.28	1.25	0.11	1.39	0.12

continued on page 62

Table 7 (continued)

Stratigraphic interval	Duration of intervals (million years)	Global structure	Total mass of sedimentary rocks	Mass, 10^{21} g				Mass per a unit of time 10^{20} g/10^6 yr	
				Volcanogenic rocks	C_{carb}	C_{org}		C_{carb}	C_{org}
1	2	3	4	5	6	7		8	9
Eocene	58–37 = 21	Continents	43.54	7.78	1.69	0.31		0.80	0.15
		Shelves and continental slopes	50.76	5.41	2.03	0.12		0.97	0.06
		Floors of oceans	21.05	0.80	1.05	0.02		0.50	0.01
		Sedimentary shell as a whole	115.35	13.99	4.77	0.45		2.27	0.22
Oligocene	37–25 = 12	Continents	21.52	2.18	0.25	0.15		0.21	0.12
		Shelves and continental slopes	34.40	4.39	0.73	0.07		0.61	0.06
		Floors of oceans	22.41	0.64	1.28	0.02		1.07	0.06
		Sedimentary shell as a whole	78.33	7.21	2.26	0.24		1.88	0.20
Miocene	25–9 = 16	Continents	45.53	5.05	0.82	0.31		0.51	0.19
		Shelves and continental slopes	65.23	5.99	1.50	0.16		0.94	0.10
		Floors of oceans	42.71	3.59	1.93	0.04		1.21	0.02
		Sedimentary shell as whole	153.47	14.63	4.25	0.51		2.66	0.31
Pliocene	9–2 = 7	Continents	12.45	2.38	0.21	0.09		0.30	0.13
		Shelves and continental slopes	22.29	1.07	0.33	0.09		0.30	0.13
		Floors of oceans	20.37	1.04	0.74	0.02		1.06	0.03
		Sedimentary shell as a whole	55.11	4.49	1.28	0.20		1.83	0.29

the life cycle and its accumulation in the Earth's crustal layers create the possibility of gaining considerable masses of free oxygen in the biosphere.

To construct Fig. 16, the above-mentioned data have been used on major reservoirs of organic carbon in the sedimentary shell of continents and oceans for the Neogäikum. These data permitted us to calculate the amount of free oxygen that has entered the atmosphere over the last 1.6×10^9 years. The influx constitutes 314×10^{20} g. Our new data helped to improve previous estimates (Ronov 1982) of oxygen volume consumed over this time in oxidizing sulphur dispersed as pyrite in sedimentary rock masses (102×10^{20} g); in oxidizing hydrogen sulphide of oceanic water to sulphates (25×10^{20} g) and iron in sedimentary rocks (41×10^{20} g). According to the estimates considering the mass of O_2 in the modern atmosphere (12×10 g), the consumption of oxygen (180×10^{20} g) is 43% less than its gain, due to residual organic carbon deposited in sediments (314×10^{20} g). The volume of oxygen equal to 134×10^{20} g proved to be imbalanced by oxidation. At present other ways of fixation of considerable oxygen masses in the sedimentary shell are unknown. The expenditure of oxygen in oxidizing Mn, V, U and other

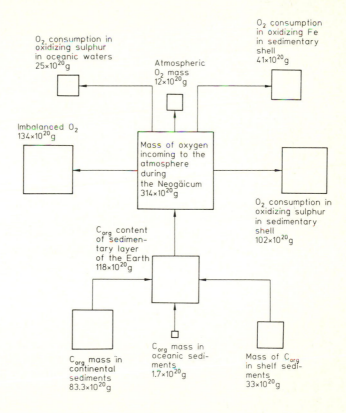

Fig. 16. The balance of oxygen incoming to the atmosphere and consumed in oxidizing processes over 1.6×10^9 years

polyvalent elements is negligible. This allows us to assume that much oxygen was consumed in oxidizing CO and other not fully oxidized gases coming into the atmosphere from the Earth's depths (CH_4, H_2 etc.). It can also be thought that some volume of oxygen (not found exactly) was spent in oxidizing polyvalent elements of igneous rocks that outcropped on the continents and the oceanic floor in different periods of the Neogäikum.

Approximate estimates of long stages of the Neogäikum show that the masses of organic carbon buried in the sedimentary shell and the corresponding masses of oxygen that entered the atmosphere increased from the Late Proterozoic to the Cenozoic (Ronov 1982). This resulted in an increasing volume of sulphate sulphur deposited in evaporites. The processes of bacterial sulphate reduction intensified in the course of diagenesis of deposited sediment series, as attested by the masses of pyrite dispersed in them. The ratio of sulphate sulphur to pyrite sulphur and of ferric oxide to ferrous oxide progressively grew and the intensity of fractionation of sulphate sulphur, and pyrite sulphur rose. Calculations show that with time the rates of accumulation of organic matter, sulphate and pyrite sulphur in the stratisphere increased, as well as the rate of oxygen production to the atmosphere. The disproportion between the production of oxygen to the atmosphere and its fixation in the sedimentary shell also grew.

The problems considered here and the regularities outlined in the carbon global balance in the Earth's sedimentary shell concern only the last stage of its development, i.e. the Phanerozoic and Late Proterozoic. For a quantitative estimate of carbon balance in the Early Precambrian, we should have to overcome great difficulties, mainly the absence of sufficiently accurate methods for obtaining global estimates, differentiated in time, of the Early Precambrian sediment masses and of masses of carbonate rocks and organic carbon buried in them.

2.3 The Dependence of Amounts of CO_2 and O_2 in the Atmosphere on Carbon Mass in Sediments

Carbon Dioxide. Time variations in the quantity of atmospheric carbon dioxide are described by the CO_2 balance equation that can be written as

$$\frac{dM_c}{dt} = A_c - B_c, \tag{4}$$

where dM_c/dt is the rate of change in carbon dioxide content, A_c the rate of carbon dioxide income to the atmosphere, and B_c the rate of its expenditure.

As noted in Chapter 1, carbon dioxide in addition to the biospheric cycle enters the atmosphere during volcanic eruptions, discharge from fissures in the Earth's crust and from different springs whose waters carry this gas to the surface.

In studying the balance of carbon dioxide it should be kept in mind that there is no direct empirical data on the rate of CO_2 production to the atmosphere during the geological past. At the same time, the rate of consumption of carbon dioxide over certain time intervals can be estimated by using the data of geochemical studies of the sedimentary shell (Ronov 1976 etc.).

In the indicated studies, the conclusion was drawn that comparatively small changes took place in the mass of continental sedimentary rocks formed over one time interval during the Phanerozoic. This implies that at this time, a greater portion of carbonate rocks resulting from the absorption of carbon dioxide was retained on the continents. In this connection, sufficiently reliable information on the consumption of carbon dioxide in the formation of carbonate rocks of the continents for the last 570 million years can be derived from data on the quantity of carbonate sediments for different time intervals.

The data in Table 6 show that the total amount of carbon dioxide consumed in forming various kinds of carbonate rocks on continents during the Phanerozoic comprises about 2×10^{23} g. This value is approximately 10^5 times greater than the present mass of atmospheric carbon dioxide. Thus if the consumption rate of CO_2 decreased or increased by half, the atmospheric CO_2 content could change by a value equal to its present quantity for the time making up 10^{-5} of the Phanerozoic duration, i.e. over several thousand years. If we assume that throughout the Phanerozoic the average amount of carbon dioxide in the atmosphere exceeded that of the present by ten times, this time would be also very short compared with the duration of geological epochs and periods.

In this connection we can conclude that for the time intervals of millions of years the value of dM_c/dt is much less than the values of A_c and B_c. Then Eq. (4) will be:

$$A_c = B_c . \tag{5}$$

Therefore the data in Section 2.2 on the mass of carbonate rocks formed over a million years in different geological epochs describe simultaneously the rates of loss and gain of atmospheric carbon dioxide.

Based on materials presented in Section 2.2, one can come to the conclusion that the rate of carbon dioxide production depended on the level of volcanic activities. To verify this assumption, comparison has been carried out between the rates of CO_2 income (expressed in grams per million years) and the quantities of volcanic rocks (V) formed on the continents over every of 28 Phanerozoic epochs.

As can be seen from Fig. 17, A_c and V are closely related. Some scatter in the points in the figure can be attributed to a limited accuracy of the A_c and V values used in constructing this graph.

For calculating changes in the atmospheric carbon dioxide content in the geological past, it is necessary to find a relationship between the mass of CO_2 and the rate of its consumption for the Earth as a whole. Because of the great variety of the processes, in the course of which atmospheric carbon dioxide is being consumed, and insufficient knowledge about them, this problem presents some difficulties.

In determining the dependence of the rate of CO_2 consumption (B_c) on its mass in the atmosphere (M_c), it should be taken into consideration that this dependence, $B_c = \varphi(M_c)$, should correspond to the following conditions: $B_c = 0$

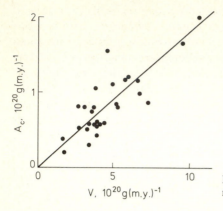

Fig. 17. The dependence of carbon dioxide income rate (A_c) on volcanic activities (v)

with $M_c = 0$ and $B'_c = \beta_c M'_c$, where B'_c and M'_c are the rate of carbon dioxide loss and carbon dioxide volume at the present epoch, and β_c the empirical coefficient.

The most simple assumption as to the nature of the dependence conforming to these conditions corresponds to the formula:

$$ B_c = \beta_c M_c , \tag{6} $$
$$ \text{where} \quad \beta_c = B'_c / M'_c . $$

This derivation of Eq. (6) is more reliable for changes in CO_2 content in the range of 0 to M'_c and for CO_2 volume slightly exceeding M'_c (the amount of carbon dioxide in the modern atmosphere). With a noticeable increase in CO_2 volume compared to M'_c, the question of the validity of Eq. (6) is less clear.

Solving this question one should consider that in the geological past, with a high carbon dioxide content, the dependence of CO_2 consumption on its concentration always seems to have the form of negative feedback. In other words, an increased atmospheric CO_2 content resulted in raising its efflux, and a decreased content, in lowering its consumption for the entire Earth.

In the absence of sufficiently strong negative feedback between B_c and M_c (and particularly in the presence of the positive relation between them), CO_2 concentration during the epochs of the greatest volcanic activities could reach very high values, which could result in an abrupt increase of surface air temperature and an approximation of the Earth's climate to the Venusian. This is discussed in detail in Chapter 3.

The comparative stability of the Earth's climate shows that within the entire range of variations in atmospheric CO_2 content its consumption increased on increasing this content. A similar conclusion can be drawn from the fact of the prolonged existence of autotrophic plants and a number of other organisms that could not have survived drastic variations in atmospheric carbon dioxide concentration.

The data in Fig. 17 are of considerable importance for confirming the hypothesis expressed by Eq. (6). As can be seen from this figure, within the wide range of variations in the level of volcanic activities, the masses of volcanic rocks are proportional to the masses of carbonates deposited during the corresponding geological epochs. It is natural to assume that the masses of volcanogenic rocks are proportional to the amount of carbon dioxide being produced into the atmosphere from the Earth's depths. Therefore, the masses of carbonate rocks are also proportional to the amount of atmospheric carbon dioxide.

In considering a certain mechanism of relation between atmospheric CO_2 content and its global consumption for carbonate formation, it should be borne in mind that the weathering of rocks on the continents is of some importance for this relation. As is known, the rate of this process depends on the carbonic acid content and the corresponding ions in the continental water. The quantity of CO_2 in these waters varies, rising in humid regions, which accelerates weathering, and increases the carrying away of alkalis of calcium, manganese and other elements from the continents into water bodies, which, combining with the dissolved CO_2 products, form limestone, dolomites and other carbonate rocks.

It might be thought that a rise in atmospheric carbon dioxide volume leads also to increased sedimentation of organic carbon, since photosynthesis of autotrophic plants usually rises with increasing atmospheric carbon dioxide partial pressure. This dependence is, however, ambiguous. To clarify this question, one can use the relationship between the rate of depositing organic carbon and the rate of accumulation of carbonate rocks (without organic carbon) for 28 epochs of the Phanerozoic (see Fig. 18). As can be seen from the figure, there is only a weak relationship between the values in question, which is explained by two reasons. The first is that the formation of organic carbon sediments is a function not only of photosynthetic productivity, but also of climatic conditions that aid in burying a greater or smaller portion of the carbon produced from organic matter created by autotrophic plants. In particular, when an arid climate spreads widely over the continents (this occurs when a considerable part of a continental area is located in the subtropical belts with high atmospheric pressure), the amount of organic carbon in sediments decreases appreciably.

The second reason for the ambiguity of the relationship can be understood from the data in Fig. 12, depicting the values of the average ratio of amount of organic carbon to amount of carbon in carbonate rocks for different epochs of the Neogäikum. From the data in this figure, the tendency to a gradual increase in this

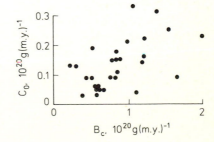

Fig. 18. The relationship between the sedimentation rates of C_{org} (C_o) and carbonate rocks (B_c)

ratio is observed. This feature can be the consequence of both a decrease in the rate of carbonate formation after the Palaeozoic (see Fig. 13) and an increase in the effectiveness of photosynthesis in more progressive types of vegetation cover. It is clear that this dependence raises the ambiguity of the relation between the rate of depositing organic carbon and the atmospheric CO_2 content.

Taking these considerations into account for calculating the amount of CO_2, in Eq. (6) one should use data on the rate of formation of carbonate rocks without including organic carbon into the volume of these rocks. At the same time, it can be mentioned that since the quantity of organic carbon is noticeably less than the amount of carbon in carbonate rocks, the results of calculating CO_2 content obtained by this method do not differ greatly from those obtained by the data on the total amount of carbonate sediments.

To calculate carbon dioxide concentration by the formula

$$M_c = \frac{M'_c}{B'_c} B_c, \qquad (7)$$

one should find the value of carbon dioxide consumption in the formation of carbonates at the present epoch, i.e. the value of B'_c.

This value can be calculated through extrapolation by using data on the amount of carbon in carbonate rocks cited in Section 2.2 that pertain to the last two epochs of the Tertiary period: the Miocene and the Pliocene. Assuming that in the mid-Miocene (17 million years ago) the rate of depositing carbon in the formation of carbonates was 0.51×10^{20}g (m.y.)$^{-1}$, and in the mid-Pliocene (5.5 million years ago) 0.30×10^{20}g (m.y.)$^{-1}$, we obtain that in the modern epoch this rate makes up about 0.20×10^{20}g (m.y.)$^{-1}$. Supposing that the total volume of the atmosphere during the Phanerozoic changed little, one can write Eq. (7) as

$$M_c = 1.5 \times 10^{-21} B_c, \qquad (8)$$

where M_c is the ratio (in per cent) of CO_2 volume to total atmospheric volume and B_c is expressed in grams of carbon in carbonate rocks formed over a million years.

The proportionality coefficient in formula (8) is taken according to the data on the rate of carbon expenditure in carbonate formation on the continents. In earlier works, the authors calculated changes in CO_2 concentration in the geological past using only the data on the formation of carbonate rocks on the continents, because there was no relevant information for the oceans. It was then assumed that at all geological epochs the rate of carbonate formation in the oceans was proportional to that on the continents. At that time it was impossible to substantiate this assumption, which made the reliability of the calculated results somewhat uncertain.

In recent years information has been obtained on the carbonate formation rate in the ocean at different geological epochs from the Late Jurassic to the Pliocene. This information can be used to verify the above-mentioned hypothesis.

Figure 19, constructed from the data given in Section 2.2, shows the formation rate of sedimentary rocks for the continents (curve 1), shelf and continental slopes (curve 2) and the oceans at considerable depths (curve 3). It is seen

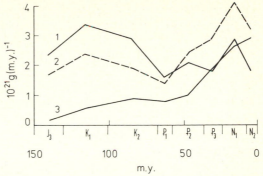

Fig. 19. The rate of sedimentary rock formation during the Late Jurassic — Pliocene (*1* continents; *2* shelves and continental slopes; *3* deep oceanic regions)

that the data for the continents oscillate, tending neither to substantial increase nor to decrease throughout the time interval under consideration. The shelf information gives evidence of a noticeably increasing rate of sedimentary rock formation at that time. This tendency is even more pronounced in the changes in sedimentary rock formation for deep oceanic regions.

Since no oceanic sediments have been discovered for the time more ancient than the Late Jurassic, it is quite evident that, in contrast to the continents, information on sedimentary rocks in the oceans covers only the final part of the Phanerozoic (about 150 million years), and within this time interval the amount of sediments diminishes the more remote the epoch.

A similar conclusion can be drawn proceeding from the data in Fig. 20, which shows the changes in the rate of carbonate rock formation on the continents, in the shelf zone and on the continental slopes, as well as in the deep ocean regions.

Thus, it is clear that the information on oceanic carbonate sediments is not complete, and the extent of its incompleteness rapidly increases the more ancient the sediments.

In order to understand the usage of the data on oceanic carbonate rocks in the calculations of the global carbon dioxide balance, we can refer to Fig. 21, which presents temporal variations in parameter *r* equal to the ratio of continental carbonates to the total amount of carbonates deposited on the continents, on shelves and continental slopes and in the deep ocean. The curve drawn through the points in the graph corresponds to the principle indicated earlier, i.e. the older

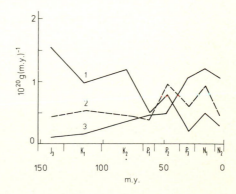

Fig. 20. Sedimentation rate of carbonate rocks during the Late Jurassic — Pliocene (*1, 2, 3* as in Fig. 19)

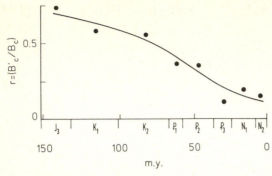

Fig. 21. The ratio r of carbonate formation rate on the continents B_c to the total rate of carbonate formation B_c during the Phanerozoic

the carbonate sediments, the greater the portion of continental rocks in the sediments. This corresponds to the loss of an ever-increasing amount of sediments formed in the modern ocean.

At the same time it is seen that almost all points in the graph are distributed close to the average curve that defines an increase in the ratio r the greater the age of the sediments. This means that making allowance for the loss of information on oceanic sediments corresponding to this average curve, the ratio of continental carbonates to their total amount will be about the same for all geological epochs. Thus, the corrected values of the global sums of carbonate sediments (oceanic sediments included) proved to be proportional to the amount of carbonates formed during different geological epochs on the continents.

This conclusion is essential for confirming the supposition that the process of carbonate rock formation is general throughout the globe and is determined to a great extend by carbon dioxide concentration, which is uniform over the entire atmosphere of the Earth. Therefore, it is possible in formula (7) to use the data on the rate of carbonate formation on the continents, considering that these data reflect changes in the rate of carbonate formation on the Earth as a whole.

Oxygen. In order to determine changes in atmospheric oxygen concentration in the geological past, we can refer to the calculations of the oxygen balance.

The changes in oxygen concentration with time dM_o/dt are defined by the equation

$$\frac{dM_o}{dt} = A_o - B_o, \tag{9}$$

where A_o and B_o are the rate of oxygen gain and loss, respectively.

In order to obtain changes in amount of oxygen, it is necessary to calculate its gain and loss for given time intervals.

As shown in Chapter 1, the rate of oxygen production to the atmosphere as a result of the activity of photosynthesizing plants is equal to the difference between the rates of oxygen production in the course of photosynthesis and its consumption on oxidation of organic matter. For the Earth as a whole, this difference at any time interval can be regarded as proportional to the amount of organic carbon that is accumulated in the sediments.

In this connection the rate of oxygen income can be found by the formula:

$$A_o = \alpha_o C_o , \tag{10}$$

where α_o is the proportionality coefficient equal to the ratio of oxygen molecular weight (O_2) to carbon molecular weight (C), i.e. 2.67; C_o is the rate of organic carbon accumulation in sediments.

It follows from the data given in Section 2.2 that in the Phanerozoic the continental sediments contained 7.1×10^{21} g of organic carbon. Taking into account that the atmosphere receives 2.67 g of oxygen per gram of deposited organic carbon, we find that, due to photosynthesis on the continents, the Phanerozoic atmosphere gained about 19×10^{21} g of oxygen.

Chapter 3 considers the question of the amount of oxygen that entered the atmosphere as a result of photosynthesis in the ocean. The conclusion is drawn that this oxygen production into the atmosphere throughout the Phanerozoic constituted about 80% of the oxygen gained from continental photosynthesizing plants, the total amount thus being approximately 34×10^{21} g.

As mentioned in Chapter 1, oxygen production to the atmosphere from all other sources is considerably less than that from photosynthesis. In particular, it may be noted that according to Brinkman's estimate (1969), throughout the Earth's history the atmosphere has received 7×10^{21} g of oxygen as a result of photodissociation of water vapour molecules. Considering that this value is probably overestimated and refers to a much longer time interval than the Phanerozoic, it is clear that the water vapour photodissociation did not play any essential role in accumulating atmospheric oxygen.

Taking account of the fact that the present atmosphere contains 1.2×10^{21} g of oxygen, it may be inferred that no less than 97% of the oxygen that entered the atmosphere during the Phanerozoic was used in the oxidation of mineral matter.

We shall further consider the possibility of using the estimated amount of oxygen absorbed in the course of oxidation of mineral matter at different past epochs in the oxygen balance calculations. In the earlier works of the authors, when such estimates were not available, the oxygen consumption was calculated based on the following considerations.

By analogy with the calculation of atmospheric carbon dioxide consumption, it should be assumed that the global oxygen expenditure on oxidizing mineral matter, i.e. the amount that leaves the biosphere's cycle, depends on its concentration in the atmosphere. With the present atmospheric oxygen content equal to M'_o, its amount spent on oxidizing mineral matter per unit time is B'_o. At the same time it is apparent that with the absence of oxygen in the atmosphere no oxygen consumption takes place, i.e. $B_o = 0$ with $M_o = 0$. Considering that B_o depends on M_o, i.e. $B_o = \psi(M_o)$, we find that the simplest form of such a relationship corresponding to the above conditions will be:

$$\text{where} \quad B_o = \beta_o M_o , \tag{11}$$
$$\beta_o = B'_o / M'_o .$$

Apart from a similar estimate for carbon dioxide, in this case there is no question that Eq. (11) should be fulfilled for high oxygen concentrations, since it is hardly possible that in the past the atmospheric oxygen content has ever much exceeded the present one. At the same time it should be borne in mind that Eq. (11) only roughly reflects the indubitable fact of growing oxygen expenditure with increase in oxygen concentration.

The data in Section 2.2 show that the amount of organic carbon that has been deposited on the continents for one million years during different Phanerozoic epochs varied from 0.03×10^{20}g to 0.33×10^{20}g. As seen in Eq. (10), this resulted in the same relative change in oxygen income to the atmosphere.

If the oxygen consumption on oxidation of mineral matter had not depended on oxygen concentration and had, for instance, been invariable, changes in oxygen content would have been very great. Let us substantiate this conclusion by a simple calculation.

By using only the data on continental organic carbon sediments (their supplementation by oceanic data does not change the results of such a calculation), we find that the average amount of oxygen used by oxidizing mineral matter throughout the Phanerozoic is about 0.35×10^{20}g per million years. At the same time, the oxygen production at different geological epochs calculated by the above-mentioned data on the rate of organic carbon accumulation in sediments varied from 0.08×10^{20} to 0.88×10^{20}g per million years. Since the duration of relevant epochs was from 7 million to 40 million years, it is easy to find that during these epochs the amount of atmospheric oxygen, its expenditure being invariable, should have decreased or increased by a magnitude of the order of 10^{21}g, which is approximately equal to the present oxygen amount in the atmosphere. If the possibility of doubling the atmospheric oxygen amount during such short time intervals compared with the Phanerozoic could not be excluded, it is evident that the conclusion as to the complete disappearance of atmospheric oxygen during the epochs when the rate of organic carbon accumulation in sediments was relatively low is clearly absurd.

It is thus certain that throughout the history of the atmosphere there has been a sufficiently strong negative feedback between the amount of atmospheric oxygen and its consumption in the oxidation of mineral matter. In earlier works the authors made a simple assumption that oxygen consumption is proportional to its volume in the atmosphere.

In order to determine more accurately the relationship $B_0(M_0)$, let us look into the concrete mechanism of oxygen consumption on oxidizing mineral matter.

As mentioned above, a considerable amount of atmospheric oxygen is being absorbed in the reactions of oxidation of sulphur and iron contained in sediments and also in the oxidation of hydrogen sulphide contained in sea water. In addition, oxygen is consumed in the oxidation of carbon monoxide and other gases that enter the atmosphere from the depths of the Earth's crust, mainly during volcanic eruptions.

Although the relative role of these forms of atmospheric oxygen consumption has been only poorly studied, there are grounds to believe that both of the indicated factors are essential for the process of atmospheric oxygen expenditure. At the same time it is evident that the oxidation rate of minerals contained in the

Earth's crust, which are to a different extent isolated from the action of atmospheric air, depends on the oxygen partial pressure, increasing with an increase in pressure. Contrary to this, the oxidation rate of gases such as carbon monoxide largely depends on the rate of carbon monoxide production into the atmosphere, and is relatively independent of oxygen pressure, while oxygen is present in the atmosphere in any substantial concentration.

Thus, if oxidation of volcanic gases had been the principal form of oxygen expenditure, that expenditure would have changed in accordance with fluctuations in volcanic activity (this will be treated below) and would have been practically independent of oxygen amount. In this case the atmospheric oxygen balance would have been unstable and the oxygen partial pressure could have varied within a wide range, which is inconsistent with the fact of the existence of complex multicellular organisms throughout the Phanerozoic.

In this connection it might be thought that the oxidation of minerals in the sedimentary layer constituted an essential part of the total oxygen expenditure on oxidizing mineral matter. Let us compare this conclusion with the inferences that can be drawn from the data presented in Section 2.2.

These data include the estimates of oxygen used by oxidation of iron and oxygen-free sulphur compounds for five long time intervals (Late Proterozoic, Early Palaeozoic, Late Palaeozoic, Mesozoic and Cenozoic).

The total oxygen expenditure for those periods proved to be about 18×10^{21} g. This, together with the present atmospheric oxygen volume, represents 57% of the amount of oxygen that entered the atmosphere over the period including the Late Proterozoic and the Phanerozoic (31.4×10^{21} g). We shall note that this calculation takes into account the data on oceanic sediments for the second half of the Mesozoic and the Cenozoic only, i.e. for a relatively short time interval. This noticeably decreases the total results of oxygen production and consumption as compared to their actual magnitudes.

Since the oxygen production appeared to be almost twice its expenditure on the oxidation of sulphurous compounds and iron, it might be thought that a considerable amount of atmospheric oxygen was consumed through oxidation of gaseous products of volcanic eruptions.

This portion of oxygen expenditure, which depends very little on oxygen concentration, was somewhat less than 43% of the total expenditure. It is highly probable that the data available on oxygen used for oxidation of minerals in a sedimentary layer are not exhaustive, and account for only the main portion of oxygen consumption.

Figure 22 presents a comparison of the average values of oxygen production (A_O) and its consumption for the oxidation of mineral substances in the sedimentary shell (B_{os}). The points in the graph correspond to the A_O and B_{os} values for five long periods of time (Late Proterozoic, Early Palaeozoic, Late Palaeozoic, Mesozoic and Cenozoic). The figure demonstrates that although the oxygen expenditure on the oxidation of sedimentary rocks is appreciably less than its production, there is a relationship between the indicated values that is close to direct proportionality.

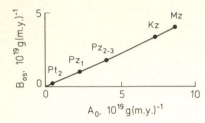

Fig. 22. The dependence of the rate of oxygen consumption on oxidation of minerals contained in sedimentary shell B_{os} on oxygen income rate A_0 for the Late Proterozoic (Pt_2), Early (P_{z_1}) and Late ($P_{z_{2-3}}$) Palaeozoic, Mesozoic (M_z) and Cenozoic (K_z)

Let us now consider two ways of calculating oxygen variations in time. The first procedure has already been used by the authors in their previous works. It is based on Eq. (9) and formulas (10) and (11), the first two relations being exact and the third approximate.

It follows that

$$\frac{dM_o}{dt} = \alpha_o C_o - \beta_o M_o. \tag{12}$$

By solving this equation for individual geological time intervals (t) we obtain

$$M_o = \frac{\alpha_o C_o}{\beta_o} + \left(M_{01} - \frac{\alpha_o C_o}{\beta_o} \right) e^{-\beta_o t}, \tag{13}$$

where M_{01} is the atmospheric oxygen concentration at the beginning of the relevant time interval.

Applying this equation and using data on organic carbon, it is possible to calculate changes in oxygen amount for different time periods of the geological past. By this technique we should assume the oxygen amount at the beginning of the entire time interval for which the calculation is carried out. For instance, in the calculation for the Phanerozoic, the assumption should be made for the oxygen volume for the beginning of the Cambrian. Afterwards we can calculate the oxygen change throughout the Early Cambrian, and the calculated result for the end of the Early Cambrian will serve as an initial value for the subsequent calculation referring to the Middle Cambrian and so on. Thus, all necessary magnitudes of parameter. M_{01} are determined in the course of the calculation itself, except for the first value assumed for the beginning of the Phanerozoic.

It can be shown that the selection of the initial value tells very little on the calculated results of changes in amount of oxygen for any geological epoch, apart from one or two, which refer to the beginning of the time interval in question. The simplest supposition in estimating M_{01} for the beginning of the Phanerozoic is that in the Early Cambrian the value of dM_o/dt was considerably smaller than the oxygen production in the biotic cycle. In this case $M_{01} = \alpha_o C_o / \beta_o$. Changing this rough estimate within reasonable range it is easy to see that these variations insignificantly affect the calculated results of oxygen amount for the epochs not later than the beginning of the Ordovician.

In order to obtain oxygen changes in time, it is necessary to find the value of parameter β_0. This parameter was not invariable and changed with time. Consideration of such changes, however, is fraught with difficulties. Therefore, in the calculation of oxygen amount we have to use an average value of parameter β_0, which can be found by the available empirical data. In particular, in earlier works the authors have used information on organic carbon in continental sediments, and the obtained value of parameter β_0, given the condition of equality between the calculated oxygen mass for the end of the Phanerozoic and its present atmospheric concentration, proved to be $3.1 \times 10^{-8} yr^{-1}$.

The accuracy of the estimates of amount of oxygen obtained in such calculations greatly depends on the completeness and reliability of the data available on organic carbon in sediments. It is therefore important to consider the data on the amount of organic carbon in oceanic sediments.

In the first calculations of oxygen changes made by the authors, when such information was not available, two different calculations based on formulas (9), (10) and (11) were made in order to understand the possible influence of that source of oxygen on atmospheric oxygen concentration. In the first of them it was assumed that the global oxygen income to the atmosphere is largely a factor of organic carbon accumulation on the continents. The second assumption allowed for the continental organic carbon deposits to be proportional to their global magnitude, although constituting only its smaller fraction. In this case $dM_0/dt \leqslant A_0$, and the atmospheric oxygen amount can be obtained by the following formula:

$$M_0 = \frac{\alpha_0 C_0}{\beta_0} . \tag{14}$$

It is clear that both of these assumptions are inaccurate. The first of them leads to overestimating the role of continental sediments and the second leads to underestimating their role in the organic carbon balance.

Figure 23 presents the values of the mean rate of the formation of organic carbon sediments on the continents (curve 1), in a shelf zone and on the continental slopes (curve 2) and in deep sea regions (curve 3). All of them are based on the data given in Section 2.2.

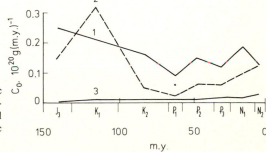

Fig. 23. Sedimentation rate of organic carbon (C_0) in the Upper Jurassic-Pliocene (1 continents; 2 shelves and continental slopes; 3 deep oceanic regions)

It follows from Figs. 19, 20 and 23 that, unlike the total mass of sedimentary rocks and the amount of carbonate sediments, the total mass of organic carbon deposited at the oceanic depths is small compared with its mass for the Earth as a whole. The amount of organic carbon deposited on the shelves and continental slopes, however, varies within a wide range and does not decline with increasing age of the sediments. Therefore, in previous work (Budyko et al. 1985), when calculating oxygen variations, in the geological past it was assumed that since the Late Jurassic the organic carbon production (A_o) can be described by the sum of the three components indicated. The oxygen production for the earlier epochs, which are not covered by the data on organic carbon sediments in the oceans, was determined by continental data, which were increased twice as much (this roughly corresponds to the average ratio of continental sediments to their global volume for the Late Jurassic-Pliocene).

The above calculation procedure of past oxygen changes can be carried out more accurately with the following considerations.

As mentioned earlier, the oxygen expenditure on oxidizing mineral substances proceeds in two principal ways, i.e. by oxidation of minerals contained in the sedimentary layer and oxidation of not fully oxidized gases that are released from the depths of the Earth's crust, carbon monoxide (CO) evidently being the most important of them. If in the former case the oxygen expenditure depends on oxygen concentration, in the latter case, in the presence of any considerable amount of oxygen, it does not depend on its concentration and is determined by the volume of not fully oxidized gases released into the atmosphere. It is natural to consider that this mass is proportional to the amount of carbon dioxide that entered the atmosphere over the same time interval. Since atmospheric carbon dioxide concentration is proportional to its gain, we may think that the second form of oxygen consumption is also proportional to carbon dioxide concentration.

Therefore, it can be considered that

$$B_o = \beta_o M_o + \gamma_o M_c , \tag{15}$$

where γ_o is the ratio of the amount of not fully oxidized gases released into the atmosphere over a certain time interval (expressed in units of oxygen mass used per unit time for oxidizing these gases) to the amount of atmospheric carbon dioxide; M_c is the amount of carbon dioxide.

Thus, the balance equation of atmospheric oxygen acquires the following form:

$$\frac{dM_o}{dt} = \alpha_o C_o - \beta_o M_o - \gamma_o M_c . \tag{16}$$

It follows from this equation by analogy with (13) that

$$M_o = \frac{\alpha_o C_o - \gamma_o M_c}{\beta_o} + \left(M_{01} - \frac{\alpha_o C_o - \gamma_o M_c}{\beta_o} \right) e^{-\beta_o t} . \tag{17}$$

In order to calculate oxygen variations by this formula, it is necessary to determine parameter γ_0. This can be done in two ways.

The observational data on the chemical composition of gases ejected through volcanic eruptions allow us to conclude that the ratio of the amount of not fully oxidized gases (expressed as oxygen amount used on oxidizing these gases) to the amount of carbon dioxide released into the atmosphere is usually confined to 1-5% (Luchitsky 1971). As seen from Table 6, about 0.37×10^{21} g of carbon dioxide per million years were used on the average in the formation of continental carbonate sediments in the Phanerozoic. Using the data from Table 7, it can be found that approximately the same amount of carbon dioxide was assimilated through the formation of carbonate sediments in the oceans, where as a result of subduction they were preserved only for the Late Mesozoic and Cenozoic. Assuming that in the Phanerozoic the mean rate of atmospheric carbon dioxide expenditure (as well as an equal mean gain in atmospheric carbon dioxide) was 0.70×10^{21} g (m.y.)$^{-1}$, we find that for the above estimate of the amount of not fully oxidized gases, $\gamma_0 M_c$ is equal from 0.7×10^{19} to 3.5×10^{19} g (m.y.)$^{-1}$. Since the mean Phanerozoic M_c is 10×10^{18} g, γ_0 proves to be 0.7-3.5 (m.y.)$^{-1}$.

Another method of estimating parameter γ_0 is based on the data on the global balance of atmospheric oxygen. It follows from Section 2.2 that the oxidation of gases entering the atmosphere from the deep Earth's layers takes up to 43% of oxygen resulting from burying a part of photosynthesizing plant biomass.

The average oxygen gain in the atmosphere for the Phanerozoic is 0.33×10^{20} g (m.y.)$^{-1}$ (after the data on continental organic carbon sediments given in Table 6). Chapter 3.1 presents the estimate of the rate of organic carbon sedimentation in the oceans, from which it follows that this rate for the Earth as a whole is 0.60×10^{20} g (m.y.)$^{-1}$. Therefore, the greatest value of $\gamma_0 M_c$ can be 0.26×10^{20} g (m.y.)$^{-1}$ and, hence, γ_0 lies within the range of 0 to 2.6 (m.y.)$^{-1}$.

It is quite probable that γ_0 is actually greater than the lower value of this range (since a considerable amount of oxygen is used in the oxidation of a part of the gases entering the atmosphere) and smaller than the upper value, because the available data on the amount of oxygen used in oxidizing minerals of the Earth's crust are evidently underestimated (see Section 2.2).

Since the value of γ_0 is estimated only roughly, the calculations of oxygen changes should be made with different magnitudes of this parameter, in order to understand how these variations influence the calculated results.

3 The Evolution of the Chemical Composition of the Atmosphere

3.1 Carbon Dioxide

Changes in Carbon Dioxide Concentration. We shall present here the calculated results of carbon dioxide fluctuations in the atmosphere for different epochs of the Phanerozoic. These have been obtained by using data on the formation rate of carbonate sediments given in Chapter 2.2 and the method of calculation described in Chapter 2.3.

Table 8 shows the values of continental carbonate sediments formed over one million years, as well as the amount of carbon dioxide and its volume percentage.

The average amount of carbon dioxide in the Phanerozoic atmosphere appeared to be 10×10^{18}g, and its average concentration 0.13% (if the total atmospheric volume was more or less invariable throughout the Phanerozoic). It can be seen that the CO_2 content of the Phanerozoic atmosphere was on the average about four times greater than that of the modern atmosphere. The range of carbon dioxide changes is from 0.03% (the present epoch) to nearly 0.3% (the Early Carboniferous), i.e. during the Phanerozoic, carbon dioxide concentration averaged over the geological epochs varied by a factor of 10.

Figure 24 shows the calculated changes in carbon dioxide concentration in the Phanerozoic (curve M_c). For comparison, the results of our previous calculation are plotted in the same figure as a dashed line (curve M'_c).

The differences between curves M_c and M'_c occurred for the following three reasons. First, the division of the Phanerozoic into geological epochs adopted in the recent calculation was more detailed. Second, the use of more accurate data on the carbonate sedimentation rate on the continents for certain geological epochs. Third, the different magnitude of the proportionality coefficient in Eq. (7) due to more precise information on the sedimentation rate of carbonates in the Miocene and Pliocene.

By comparing curves M_c and M'_c it can be seen that, except for the Cambrian, they reflect almost identical relative fluctuations in carbon dioxide over time. The earlier data on amounts of carbon dioxide are somewhat 'higher than those obtained recently. This discrepancy is explained by the different values of proportionality coefficient in formula (7), which substantially depend on relatively small variations in the rate of carbonate formation in the Miocene and

Table 8. The amount of atmospheric carbon dioxide in the Phanerozoic

			1	2	3
Stratigraphic interval		Duration of intervals, m.y.	Carbonate sediments, 10^{20}g (m.y.)$^{-1}$	Carbon dioxide, 10^{18}g	Carbon dioxide, %
E. Cambrian	$\in_1$	570−545 = 25	0.43	4.9	0.064
M. Cambrian	$\in_2$	545−520 = 25	0.58	6.7	0.087
L. Cambrian	$\in_3$	520−490 = 30	0.38	4.4	0.057
E. Ordovician	O_1	490−475 = 15	1.17	13.5	0.175
M. Ordovician	O_2	475−450 = 25	0.86	9.9	0.129
L. Ordovician	O_3	450−435 = 15	0.85	9.8	0.127
E. Silurian	S_1	435−415 = 20	0.58	6.7	0.087
L. Silurian	S_2	415−402 = 13	0.57	6.6	0.086
E. Devonian	D_1	402−378 = 24	0.58	6.7	0.087
M. Devonian	D_2	378−362 = 16	0.82	9.5	0.123
L. Devonian	D_3	362−346 = 16	1.20	13.9	0.180
E. Carboniferous	C_1	346−322 = 24	1.99	23.0	0.299
M.-L. Carboniferous	C_{2-3}	322−282 = 40	0.81	9.4	0.122
E. Permian	P_1	282−257 = 25	1.65	19.0	0.247
L. Permian	P_2	257−236 = 21	0.59	6.8	0.088
E. Triassic	T_1	236−221 = 15	0.60	6.9	0.090
M. Triassic	T_2	221−211 = 10	1.11	12.8	0.166
L. Triassic	T_3	211−186 = 25	0.75	8.6	0.112
E. Jurassic	J_1	186−168 = 18	0.80	9.2	0.120
M. Jurassic	J_2	168−153 = 15	1.05	12.2	0.158
L. Jurassic	J_3	153−133 = 20	1.54	17.8	0.231
E. Cretaceous	K_1	133−101 = 32	0.99	11.4	0.148
L. Cretaceous	K_2	101−67 = 34	1.19	13.7	0.178
Palaeocene	P_1	67−58 = 9	0.51	5.8	0.076
Eocene	P_2	58−37 = 21	0.80	9.2	0.120
Oligocene	P_3	37−25 = 12	0.21	2.5	0.032
Miocene	N_1	25−9 = 16	0.51	5.8	0.076
Pliocene	N_2	9−2 = 7	0.30	3.5	0.045

Pliocene. This dependence is a shortcoming of the calculation procedure and leads to lower accuracy in determining the absolute values of carbon dioxide fluctuations in the geological past; but at the same time it does not affect the calculation of relative changes in carbon dioxide, whose accuracy is much higher than the accuracy of determining its absolute values.

Considering the results of the recent calculation (curve M_c in Fig. 24), it can be seen that during the Phanerozoic six maxima of carbon dioxide concentration took place which noticeably exceeded the average level of CO_2 concentration. The first rise occurred in the Early Ordovician, the second started in the Devonian, and in the Early Carboniferous the carbon dioxide concentration was highest for the entire Phanerozoic. The third rise, which was also the second greatest, took place in the Early Permian, the fourth in the Middle Triassic, the fifth in the Late

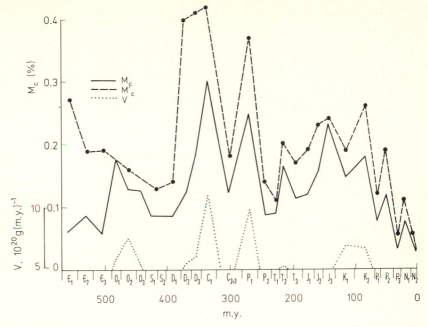

Fig. 24. Changes in carbon dioxide concentration (M_c, M_c') and the rate of formation of volcanic rocks (V) during the Phanerozoic

Jurassic, and the sixth in the Late Cretaceous. The last two maxima overlap a noticeable increase in the level of carbon dioxide concentration that embraced most of the Jurassic and the entire Cretaceous.

The causes of these increases can be easily understood when considering fluctuations in the level of volcanic activity. It can be found from data in Chapter 2.2 that the average Phanerozoic rate of volcanic rock formation is 4.8×10^{20} g (m.y.)$^{-1}$. Figure 24 shows the formation rate of volcanic rocks referring to the geological epochs when volcanic activity was above average.

It can be seen that the time of occurrence of the first four maxima in volcanic activity accords well with the maxima of carbon dioxide concentration. The last maximum of vulcanicity that embraced the entire Cretaceous also corresponds to a noticeable rise in carbon dioxide concentration during the Late Jurassic and Cretaceous.

In order to understand the present natural conditions, attention should be paid to a gradual lowering of carbon dioxide concentration from the second half of the Cretaceous to the present time, when this concentration reached the lowest level for the entire Phanerozoic. The calculated results of carbon dioxide concentration and the information on the intensity of volcanism for this time interval are given in Fig. 25. The figure shows that the process of decreasing CO_2 concentration that took place since the Late Cretaceous was not continuous.

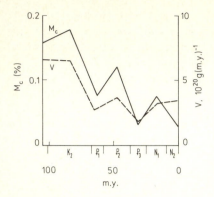

Fig. 25. Changes in carbon dioxide concentration (M_C) and the rate of formation of volcanic rocks (V) in the Late Cretaceous-Cenozoic

Even with limited time resolution of the available data, it can be seen that throughout this time interval carbon dioxide concentration appreciably increased twice, in the Eocene and in the Miocene, following the CO_2 maximum in the Late Cretaceous. However, each of these rises in carbon dioxide level was smaller than the previous one, i.e. the CO_2 level was gradually decreasing. At the same time the successive minima in carbon dioxide concentration since the Early Cretaceous (including the minima of Palaeocene and Oligocene) also reflect the lowering of the carbon dioxide level. It should be noted that fluctuations in atmospheric carbon dioxide concentration follow the changes in volcanic activity (the Pliocene partially excluded) given in Fig. 25. Therefore, it seems evident that the tendency to decreasing CO_2 concentration, which prevailed throughout the time interval in question, is explained by the process of decreasing volcanic activity.

Turning to the causes of fluctuations in volcanic activity in the geological past, we note that throughout the Phanerozoic there were five considerable maxima in volcanic activity that took place on the average about every 100 million years (see Fig. 24). This conclusion can be compared to the results of investigations carried out by Tikhonov et al. (1969). They theoretically studied the thermal convection in the mantle and concluded that this process was rhythmical: the upheaval of heated substance to the Earth's surface occurred about every 100 million years. The coincidence between these time intervals allows the supposition that the primary cause of carbon dioxide fluctuations in the atmosphere was the process of uneven degassing of the mantle. This was a result of periodic uplifting of heated substance to the Earth's surface by thermal convection.

It can be considered that this process is characterized not only by the main rhythm mentioned above but by shorter rhythms, which can be seen in Fig. 25. One can notice increases in the level of volcanic activity whose amplitude is smaller and which are divided by time periods of about 30 million years.

It is sometimes supposed that, in addition to rhythmical fluctuations in volcanic activity, there has been a general tendency to its decrease throughout the Earth's history. This tendency is assumed to be explained by a gradual decrease in the amount of preserved long-lived radioactive isotopes of the elements which, on disintegration, heat the Earth's deeper layers. This supposition would account for the presence of the great amount of carbon dioxide in the ancient atmosphere, which much exceeded the atmospheric CO_2 content in the Phanerozoic.

There are certain difficulties in estimating the CO_2 content of the Precambrian atmosphere by the geochemical evidence available. The studies mentioned in Chapter 2 show that in the Late Proterozoic (1600-570 million years ago) the sedimentation rate was 0.26×10^{21}g (m.y.)$^{-1}$. This value is much lower than that for the Phanerozoic, which was 2.3×10^{21}g (m.y.)$^{-1}$. Thus, it is evident that only a small portion of sedimentary rocks of the Late Proterozoic have been preserved until the present.

The relative amount of carbon in the Late Proterozoic carbonate rocks, equal to 4.72% of the sedimentary shell, is greater than that for the Phanerozoic, which is 4.08%. Proceeding from the idea that atmospheric CO_2 content is associated with the amount of carbonate sediments, it can be concluded that the average CO_2 content of the Late Proterozoic atmosphere was 16% greater than that of the Phanerozoic. This corresponds to a carbon dioxide concentration of 0.15% (with the present atmospheric volume). This conclusion has been made assuming that the Late Proterozoic portion of the present carbonate rocks is close to the total amount of preserved sedimentary rocks. Since this question is not very clear, the estimation of the amount of carbon dioxide in the Precambrian atmosphere will be treated further in this section.

Let us now compare our results with the conclusions made by other authors, who have studied changes in atmospheric carbon dioxide in the geological past. As already mentioned in Chapter 1, it has long been supposed that the atmospheric CO_2 content decreased throughout the history of the atmosphere and, in particular, during the Phanerozoic. Our results have shown that this supposition holds true mainly for the close of the Phanerozoic, which lasted no more than one hundred million years.

The available estimates of the range of carbon dioxide fluctuations in the Phanerozoic atmosphere were mostly based on qualitative considerations and were rather conjectural than scientifically confirmed. However, some of these estimates are more or less correct. For example, Rutten (1971) assumed that maximum carbon dioxide content in the Phanerozoic was five times greater than its present mass. This estimate does not differ greatly from our result. It is also worth noting the suggestion of Garrels (1975), who estimated the range of CO_2 fluctuations in the atmosphere during the Phanerozoic as from 0.01 to 0.09%. Although the minimum and maximum values of this range are three times lower than our results, the range of relative fluctuations in carbon dioxide has been evaluated correctly.

There are studies which state, without sufficient ground, that at the beginning of the Phanerozoic carbon dioxide was one of the main constituents of the atmosphere. The data of the next section show that such a supposition is incompatible with the information available on climates in the geological past. It also disagrees with the findings on plant and animal evolution.

As mentioned in Chapter 1, a detailed quantitative calculation of variations in the atmospheric chemical composition throughout the Earth's history has been made by Hart (1978). However, in calculating carbon dioxide changes Hart has not employed data on sedimentary composition and therefore some of his results proved to be unrealistic. In particular, Hart thought that during the Phanerozoic carbonates did not form by the degassing of the Earth's deeper layers and consequently, he concluded that the CO_2 content of the atmosphere did not change at that time. Since actually the basic portion of the existent carbonates was formed from the gases that entered the atmosphere in the Phanerozoic, his conclusion appeared to be wrong. Of prime importance in verifying the reliability of the data presented here is information on climatic conditions of relevant epochs obtained in palaeoclimatic studies, which is considered below.

Carbon Dioxide and Climate. The central question in the study of the atmosphere's history is the influence of carbon dioxide changes on climate. This is explained by several reasons. The first is associated with the change in global climate occurring at present as a result of increasing atmospheric carbon dioxide content due to man's economic activities. The evaluation of future climatic changes requires data on the chemical composition of the atmosphere in the geological past as well as information on ancient climates, which were dependent on the CO_2 content of the atmosphere. The results of studying the evolution of the atmosphere can therefore be used when solving vital practical problems. This question will be covered in Section 3.3.

The second reason is that without understanding CO_2 climatic effects, it is impossible to perceive why the climate has changed relatively little throughout billions of years and how life could have survived on Earth. These questions are also treated in Section 3.3.

Finally, the third reason is associated with the possibility of using data on the climatic effect of CO_2 changes in the past in checking the calculations of fluctuations in atmospheric CO_2 concentration in the geological past. This can be of greater importance in the calculation of carbon dioxide concentrations compared to the calculations of the amount of oxygen. It can be explained by the following. In the calculations of O_2 income it is possible to rely on empirical data; but the absence of information directly characterizing the CO_2 gain in the atmosphere during different geological epochs hampers the verification of theoretical calculations of carbon dioxide concentration.

As mentioned in Chapter 1, the idea of climatic effects caused by carbon dioxide changes in the geological past was expressed at the turn of the 19th and 20th centuries by Arrhenius (1896, 1903) and Chamberlin (1897, 1898, 1899). These scientists presumed that CO_2 fluctuations could cause Quaternary glaciations. Arrhenius was also sure that warm climatic conditions in the past resulted from an increase in atmospheric CO_2 content.

In studying the effects of CO_2 on climate it is important to understand the dependence of the mean surface air temperature on the amount of carbon dioxide in the atmosphere. Callendar (1938) found on the basis of very simplified calculations that under conditions of increasing CO_2 content, the air temperature increases rapidly with low CO_2 concentrations and more slowly with high CO_2 content. This conclusion was confirmed by a detailed study of Manabe and Wetherald (1967) and also by Augustsson and Ramanathan (1977), who showed that the dependence of surface air temperature on CO_2 concentration is logarithmic.

To estimate quantitatively temperature variations on the basis of this dependence, it is necessary to know the value of temperature increase with a given increase in CO_2 concentration. For this purpose, parameter ΔT_c is ordinarily used, which corresponds to temperature increase with a doubling of CO_2 concentration from 0.03% to 0.06%.

Although in the earlier studies this parameter used to be determined by very approximate methods, later some of the earlier values of ΔT_c proved to be close to the present estimates. For example, it follows from Arrhenius' data (1908) that $\Delta T_c = 4°C$. According to Grigoriev (1936), it follows that $\Delta T_c = 3.1°C$. At the same time, many of the earlier results and even some of the modern estimates of this parameter are not realistic, and this hampers the study of climatic effects of carbon dioxide.

Today, in order to determine parameter ΔT_c, investigators use the most detailed climatic models as well as empirical methods based on the dependence of present climatic changes on the growth of CO_2 concentration in the atmosphere.

The most important results of such studies are published in a review by Budyko et al. (1983) and presented in Table 9.

As can be seen from Table 9, all the estimates of parameter ΔT_c differ comparatively little from each other. The average value of this parameter is close to 3°C, the individual estimates deviating from this magnitude on the average by 20%.

Table 9. The change in air temperature on doubling CO_2 concentration

Simplified climatic models	General circulation models	Data on the present climate change
1. 2.4°C	8. 2.9°C	16. 3.3°C
2. 2.5-3.5°C	9. 3.9°C	17. 2.0-3.0°C
3. 2.0-3.2°C	10. 3.5°C	18. 2.1-4.2°C
4. 3.3°C	11. 2.0°C	
5. 4.4°C	12. 2.0°C	
6. 3.2°C	13. 3.0°C	
7. 3.5°C	14. 2.4°C	
	15. 2.5°C	

Estimates 1, 8, 13 — Manabe and Wetherald (1967, 1975, 1980); 3 — Augustsson and Ramanathan (1977); 4 — Ramanathan et al. (1979); 5 — Kondratiev and Moskalenko (1980); 6 — Hameed et al. (1980); 7 — Mokhov (1981); 9, 10 — Hansen et al. (1979); 11 — Manabe and Stouffer (1980); 12 — Dymnikov et al. (1980); 14 — Wetherald and Manabe (1981); 15 — Bryan et al. (1982); 2, 16 — Budyko (1974, 1977b); 17, 18 — Vinnikov and Groisman (1981, 1982).

The average value coincides with the estimate of parameter ΔT_c given in the Reports of the U.S. National Academy of Sciences (Carbon . . . . . 1979, 1982), in the Report of the Soviet-American Meeting of Experts (Climatic Effects of Increased Atmospheric Carbon Dioxide 1982) and in some other collective works in this field.

Using this information, it is possible to find the mean global temperature variations in the geological past. Calculation of this nature was carried out earlier by Budyko (1981) on the basis of the following considerations.

It follows from climatic theory that the mean air temperature of the global surface depends mostly on the following three factors: the atmospheric CO_2 content, the solar constant (i.e. the income of solar radiation per unit area of the outer atmosphere with an average distance between the Earth and the Sun) and the reflectivity (the albedo) of the Earth for solar radiation.

The information given in the previous section shows that the CO_2 content of the atmosphere repeatedly increased and decreased throughout the Phanerozoic. Apparently, this stimulated a relevant increase or decrease in the air temperature. Solar luminosity is supposed to have increased continually throughout the history of the Sun, at a rate of 5% per billion years (Newman and Rood 1977). Other things being equal, this factor should have caused an increase in surface air temperature.

Throughout the history of the atmosphere, variations in the Earth's albedo could have been caused by two factors. The first is changes in cloudiness, and these probably did not greatly influence the mean surface air temperature, because with an increase in cloudiness both the amount of short-wave radiation absorbed by the Earth and the energy loss by outgoing long-wave radiation decrease. Since these effects roughly compensate for each other, changes in cloudiness are presumed to influence the mean surface air temperature only slightly. Another factor could have been of greater importance, i.e. past changes in the Earth's surface albedo, which affected the albedo of the Earth-atmosphere system and consequently, the Earth's albedo as a whole. Since the Earth's surface albedo depends on the area of water bodies, the Earth's albedo was somewhat decreased at the beginning of the Phanerozoic because of the relatively large area of water bodies (the albedo of which is lower than that of the continents). With subsequent changes in this area (see Klige 1980), the Earth's albedo changed accordingly.

At the same time, the Earth's albedo was increasing with the development of more or less large glaciations, because the reflectivity of snow and ice is generally high. Besides, the Earth's albedo was influenced to some extent by the continental forests, whose albedo is usually smaller than that of forestless territories.

The effect of the two latter factors on the mean albedo can be found only approximately. This, however, makes little change in the accuracy of temperature calculations, since the mean air temperature variations were generally most dependent on the first two factors.

Moreover, the mean temperature in the geological past could have been influenced by the location of the continents and oceans, which undoubtedly greatly affected the atmospheric circulation and sea currents and as a result, the position of the climatic zones. Some calculations based on climatic theory yielded the conclusion that with unvarying albedo of the Earth's surface, differences in the motions of air and water masses change the mean global temperature relatively little.

Although this question requires further investigation, there are now no grounds to think that the effect of the location of the continents and oceans on mean temperature is comparable with the effect of these basic factors.

Table 10 gives the calculated difference between the air temperature in the geological epochs of the Phanerozoic and the present air temperature. Apart from the calculation carried out in previous work (Budyko 1981), we have now used the data on CO_2 concentration in the past from this work (Table 8) and assumed the value of parameter ΔT_c equal to 3°C.

Table 10 presents the values of CO_2 concentration used in the calculation, the ratio of the difference between the amount of solar radiation, present and past, to the contemporary value of the solar constant ($\Delta S/S$) and similar differences for

Table 10. The difference between mean air temperatures in the geological past and the present epoch

Time interval		Duration of interval, m.y.	$M_c\%$	$\Delta S/S$	$-\Delta\alpha$	$\Delta T°C$
E. Cambrian	ϵ_1	570— 545 = 25	0.064	0.028	0.018	2.9
M. Cambrian	ϵ_2	545—520 = 25	0.087	0.026	0.015	3.9
L. Cambrian	ϵ_3	520—490 = 30	0.057	0.025	0.015	2.3
E. Ordovician	O_1	490—475 = 15	0.175	0.024	0.017	7.7
M. Ordovician	O_2	475—450 = 25	0.129	0.023	0.017	6.5
L. Ordovician	O_3	450—435 = 15	0.127	0.022	0.017	6.6
E. Silurian	S_1	435—415 = 20	0.087	0.021	0.015	4.7
L. Silurian	S_2	415—402 = 13	0.086	0.020	0.015	4.8
E. Devonian	D_1	402—378 = 24	0.087	0.019	0.016	5.1
M. Devonian	D_2	378—362 = 16	0.123	0.018	0.015	6.6
L. Devonian	D_3	362—346 = 16	0.180	0.018	0.015	8.3
E. Carboniferous	C_1	346—322 = 24	0.299	0.016	0.016	11.0
M.-L. Carboniferous	C_{2-3}	322—282 = 40	0.122	0.015	0.005	5.0
E. Permian	P_1	282—257 = 25	0.247	0.013	0.004	8.0
L. Permian	P_2	257—236 = 21	0.088	0.012	0.016	6.2
E. Triassic	T_1	236—221 = 15	0.090	0.011	0.015	6.2
M. Triassic	T_2	221—211 = 10	0.166	0.011	0.014	8.7
L. Triassic	T_3	211—186 = 25	0.112	0.010	0.014	7.1
E. Jurassic	J_1	186—168 = 18	0.120	0.009	0.015	7.7
M. Jurassic	J_2	168—153 = 15	0.158	0.008	0.016	9.3
L. Jurassic	J_3	153—133 = 20	0.231	0.007	0.015	10.9
E. Cretaceous	K_1	133—101 = 32	0.148	0.006	0.015	9.2
L. Cretaceous	K_2	101—67 = 34	0.178	0.004	0.016	10.4
Palaeocene	P_1	67—58 = 9	0.076	0.003	0.015	6.7
Eocene	P_2	58—37 = 21	0.120	0.002	0.014	8.5
Oligocene	P_3	37—25 = 12	0.032	0.002	0.014	2.8
Miocene	N_1	25—9 = 16	0.076	0.001	0.012	6.4
Pliocene	N_2	9—2 = 7	0.045	0	0.008	3.4

the Earth's albedo $\Delta\alpha$ expressed in fractions of a unit. The value of the mean global air temperature change compared to the present epoch ΔT was assumed to be equal to the sum of three temperature differences dependent on changes in carbon dioxide, solar radiation and albedo. According to the dependencies found earlier (Budyko 1980), it was supposed that 1% increase in the solar radiation with invariable albedo raises the mean air temperature by 1.4°C, and an increase in the Earth's albedo by 0.01 lowers the temperature by 2°C.

Taking into account the fact that at present the mean surface air temperature is equal to 15°C, it is possible to find from Table 10 the values of temperature T for each geological epoch presented in Fig. 26. As can be seen from the figure, throughout the Phanerozoic up to the Pliocene the mean temperature was 2.8°-11.0°C higher than at present, i.e. it was 18°-26°C. At the end of the Phanerozoic (in the Oligocene, in the Pliocene and afterwards in the Quaternary period) noticeable air temperature decreases took place. Let us note that during the Quaternary period there were considerable temperature fluctuations, which are impossible to present in Fig. 26, since they were relatively short-term. During the coldest epochs of the Pleistocene corresponding to the development of large glaciations, the mean surface air temperature dropped by approximately 5°C compared to the present.

The calculations by the climatic theory show that during the Phanerozoic epochs with warmer climate the air temperature in the tropical latitudes increased comparatively little, while in the middle and particularly in high latitudes it was much higher than at present. As a result, the polar climate was usually relatively warm, which precluded the existence of polar ice. This is also confirmed by palaeogeographic studies, which in particular have established that in the Mesozoic and during the warm Cenozoic epochs forests consisting of evergreen trees grew in the high latitudes of the Northern Hemisphere.

The calculations of temperature variations in the Phanerozoic have shown that these variations were mostly dependent on carbon dioxide fluctuations. Therefore, if we have temperature data for the geological past obtained by independent methods, it is possible to verify the estimates of atmospheric carbon dioxide content given in Table 8.

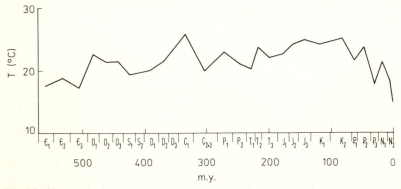

Fig. 26. Changes in the mean air temperature (T) in the Phanerozoic

As mentioned in Chapter 1, palaeoclimatic studies are based on information about the types of sedimentary rock formations, on geomorphological data and on the evidence of fossil flora and fauna. These studies take into account the relationships between the lithogenesis and climatic factors, as well as the influence of climatic conditions on the geographical distribution of living organisms and particularly on the distribution of plants. The information on fossil spore and pollen is very important, because it allows the determination of the composition of the vegetation cover in the region under consideration. This approach yields more reliable results for the not so remote past, when plants differed little from modern forms and when the climatic conditions affected the plants' dissemination in just the same way as today.

The most usual method applied in modern palaeoclimatogy for determining palaeotemperature is by the isotope ^{18}O content of the fossil remains of aquatic animals. It has been found that the $^{18}O/^{16}O$ ratio in the shells of molluscs and other remains of marine organisms depends on several factors, including the environmental temperature under which these organisms existed.

More reliable information on the climates of the geological past occurs for the Late Phanerozoic, mainly for the Cenozoic. In addition, for this time there is most convincing evidence on the amount of carbonate rocks in sediments, which is used when calculating CO_2 concentration. It should also be taken into account that the location of the continents and the oceans during the Cenozoic (and especially in the Neogene) approached the modern one more closely than earlier epochs. Therefore, the Cenozoic data and partly the data for the second half of the Mesozoic are most important when checking the calculated results of CO_2 concentration.

The most detailed picture of climatic changes in the Phanerozoic was drawn by Sinitsyn (1965, 1966, 1967, 1970, 1976), who summarized the data on lithogenesis, floras and faunas of the geological past. Using these data, Sinitsyn constructed a number of palaeoclimatic maps for different periods and epochs of the Phanerozoic. These maps cover the territories of Europe, extratropical Asia and North Africa. Although new extensive information on Phanerozoic climates has recently appeared, based on palaeotemperature measurements, the present studies generally cover limited areas and do not contain palaeoclimatic maps for the entire Earth or its different parts. Therefore, Sinitsyn's maps are important in the calculation of mean global temperature variations throughout the Phanerozoic.

These maps can be used for calculating the differences between the mean annual air temperature for various geological epochs and the present temperature for the latitudinal zone of 30°-80°N (ΔT_{30-80}). Extrapolating the changes obtained in mean latitudinal temperature differences up to the equator and the pole, it is possible to find the mean air temperature differences for the Northern Hemisphere (ΔT_{0-90}). The values of ΔT_{30-80} and ΔT_{0-90} for the end of the Mesozoic and for some epochs of the Tertiary are presented in Table 11. This table also includes the calculated data from Table 10 on the mean global air temperature differences for the same time intervals, designated by ΔT.

Table 11. The difference between mean air temperatures in the geological past and the present time

Time interval	$\Delta T_{30-80}\,°C$	$\Delta T_{0-90}\,°C$	$\Delta T°C$
Cretaceous	17.4	11.0	9.8
Palaeocene-Eocene	15.2	8.2	7.9
Miocene	11.8	6.0	6.4
Pliocene	9.2	4.8	3.4

As can be seen in Table 11, the air temperature differences for four epochs obtained by calculation and by Sinitsyn's maps deviate from each other on the average by less than 1°C. This value represents a small portion of the range of air temperature changes. Such a result proves the reliability of the data on atmospheric carbon dioxide in the geological past given in this study.

The same conclusion can be made from Fig. 27, which shows the magnitudes of $\Delta T'$ as dots corresponding to carbon dioxide concentrations during the geological epochs closing the Phanerozoic, derived from Table 8. The values of $\Delta T' = \Delta T - \Delta T_s - \Delta T_\alpha$ (where ΔT_s is the temperature change because of the solar constant increase and ΔT_α is the temperature change because of albedo fluctuations) are temperature variations produced by CO_2 fluctuations determined by empirical data. The curve in the figure corresponds to the logarithmic dependence of $\Delta T'$ on CO_2 concentration with $\Delta T_c = 3°C$. The curve agrees well with the dots on the graph, which confirms the reliability of the calculated values for carbon dioxide concentration.

The calculated results of palaeotemperatures can also be compared with empirical data on temperature variations in certain geographical regions. Figure 28 presents temperature differences between Cenozoic epochs and the present time for the following three regions: the west of the USA (curve 1), the mid-latitude oceans in the Southern Hemisphere (curve 2), the southern North Sea (curve 3). The first curve is based on palynological evidence (Axelrod and Baily 1969), the second (Shackleton and Kennett 1975) and the third (Buchardt 1978) are based on the isotope content of marine organisms. For comparison, the figure shows the earlier data on CO_2 changes during the Cenozoic. It can be seen that the

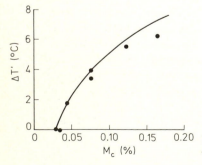

Fig. 27. The dependence of the mean air temperature ($\Delta T'$) on carbon dioxide concentration M_c

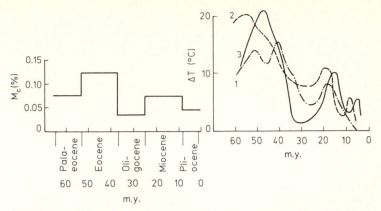

Fig. 28. Changes in carbon dioxide concentration (M_c) and mean air temperature (ΔT) during the Cenozoic (*1* west of the USA; *2* mid-latitude oceans in the Southern Hemisphere; *3* the southern North Sea)

independent data on temperature variations agree well with those on carbon dioxide fluctuations. For instance, all temperature curves exhibit a warming trend that occurred at the end of the Palaeocene-Eocene. One can also see an abrupt cooling in the Oligocene, a temperature maximum in the Miocene and a cooling in the Pliocene compared to the Miocene. Certain differences in the indicated temperature maxima and minima obtained by the data derived from different studies are explained by differences in time scales used in these studies.

In addition to the principal features of climatic changes during the Tertiary, Fig. 28 also reveals some characteristics of the changes in thermal regime during time intervals that are shorter than the geological epochs. Therefore, it is impossible to compare these changes with fluctuations in carbon dioxide concentration.

Our results of Phanerozoic changes in amount of carbon dioxide have been compared to the data on the past climatic conditions in a number of studies (Tenyakov and Yasamanov 1981; Frakes 1984, etc.), satisfactory agreement being found between climatic changes and fluctuations in carbon dioxide concentration.

If the empirical data on the mean global temperature of the Cenozoic epochs agree well with the results calculated from the data on carbon dioxide concentration, it is evident that through palaeotemperature data we can solve the inverse problem, i.e. determine the sensitivity of the Earth's thermal regime to changes in carbon dioxide concentration. An example of such calculation is presented in Fig. 29. The values of $\Delta T'$ are found by the above method and then compared to the values of CO_2 concentrations expressed on a logarithmic scale. Considering the angle of inclination of the straight line drawn through six points, it is possible to find that $\Delta T_c = 3.0°C$.

According to the earlier calculations of the same kind, the value of parameter ΔT_c was found to be 2.8°-3.5°C (Budyko 1977b, 1980). These values agree well with independent estimates of this parameter given in Table 9.

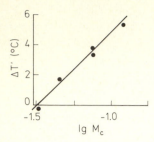

Fig. 29. Determination of parameter ΔT_c

In their previous works, the authors have also presented other grounds to prove the reliability of calculating changes in atmospheric carbon dioxide during the Phanerozoic. Among them, for example, was the above-mentioned good agreement between the epochs of maximum CO_2 concentrations and high volcanic activity. At the same time, the calculated average carbon dioxide concentration for the Phanerozoic (0.10-0.15%) proved to approach the CO_2 concentration that is optimum for photosynthesis. This confirmed the idea that modern autotrophic plants were adapted to the chemical composition of the atmosphere, which was richer by far in carbon dioxide than in our epoch. We would also like to mention the fact that the calculated range of relative variations in carbon dioxide concentration coincides with some of the earlier suppositions as to how much this concentration could vary based on the principles of sedimentary rock formation.

It has recently been concluded as a result of studying the chemical composition of air bubbles in ice cores that carbon dioxide concentration changed noticeably during the Quaternary glaciations (Neftel et al. 1982). The CO_2 concentration decreased during the glacial epochs and increased during the warm interglacials. These changes seem to have had considerable influence on climatic fluctuations and the state of ice sheets in the Quaternary. Since such fluctuations in carbon dioxide concentration have been revealed for the time intervals of thousands of years, it is quite probable that similar fluctuations also took place during longer time intervals extending for millions of years.

However, these considerations are less important in confirming the reliability of calculated Phanerozoic carbon dioxide fluctuations than the conclusion that CO_2 fluctuations corresponded to variations in the atmospheric thermal regime.

In conclusion to this section, let us dwell on a more difficult question, the amount of carbon dioxide in the Precambrian atmosphere. It has already been mentioned that the data on the relative amount of carbonate rocks in the Late Proterozoic sediments show that the average carbon dioxide concentration at that time could have been 16% greater than that of the Phanerozoic. With invariable atmospheric volume this would have corresponded to a CO_2 concentration equal to 0.15%. Although this estimate is very rough, it deserves attention, because it allows us to make a reasonable conclusion about climatic conditions of the Late Proterozoic.

As compared to the present epoch, the indicated amount of carbon dioxide should have increased the mean air temperature by 7°C. At the same time, at the beginning of the Late Proterozoic, the solar radiation was 8% lower than at present and at the end of this time interval it was 3% lower. This corresponds to a decrease in mean temperature by 11°C and 4°C, respectively. On the average this is 7.5°C, the value which is close to the indicated temperature increase. As a result, the Late Proterozoic climate on the whole appears not to differ greatly from contemporary climatic conditions. It is possible that throughout the Earth's history climatic conditions averaged over long time intervals have varied relatively little.

At the same time during the Late Proterozoic, like in the Phanerozoic considerable fluctuations in volcanic activity undoubtedly took place. In the epochs of high volcanic activity the amount of atmospheric carbon dioxide increased. When the level of volcanism was low, the CO_2 concentration decreased. Consequently, the air temperature also changed.

The epochs of major glaciations in the second half of the Late Proterozoic, approximately dated as 940, 770 and 615 million years ago (Frakes 1979), could have corresponded to the periods with attenuated vulcanicity.

This supposition agrees well with the foregoing estimates of the Late Proterozoic thermal regime. If the thermal regime of that time roughly corresponded to the present climatic conditions, then the Earth's albedo (and, hence, the mean surface air temperature) should have depended very much on the area of ice that partly covered the continents and the oceans. When, as a result of the continental drift, the oceans retreated from one or both of the poles, the ice area increased during the time of low level of volcanic activity. This led to an increase in the albedo and resulted in a further temperature decrease. According to climatic models, such a "positive feedback" could have caused large glaciations.

The history of the Phanerozoic climate shows that with the mean air temperature exceeding the present one by more than 5°C, large glaciations could hardly have arisen. As mentioned above, during the major Quaternary glaciations the mean air temperature decreased by approximately 5°C compared with the present epoch. In addition to the Quaternary glaciations developing at a mean air temperature that was very low for the Phanerozoic, during the last 600 million years there has been one more large glaciation, at the end of the Carboniferous and the beginning of the Permian. We have the mean temperature data only for the epochs (Middle-Late Carboniferous and Early Permian), which were much longer than the duration of the glacial intervals. Each of these epochs witnessed no glaciations for many millions of years. Therefore it is probable that at the end of the first of these epochs and at the beginning of the next, the air temperature was much lower than the data presented in Table 10. At the same time during the Early Permian the air temperature increased considerably because of an abrupt intensification of volcanic activity and very high CO_2 concentration, which created the conditions for the retreat of this extensive glaciation.

In conclusion let us discuss Fig. 30 (Lapenis 1984, with additions), which shows the results of several independent calculations of carbon dioxide changes

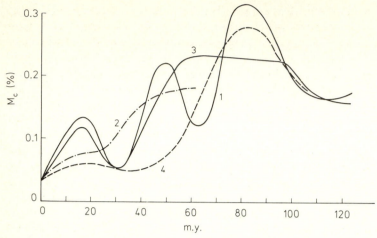

Fig. 30. Changes in carbon dioxide concentration at the end of the Phanerozoic according to different data

in the atmosphere for the last 110 million years. Curve 1 is constructed by the data on the relationship between the CO_2 content and the formation rate of carbonate sediments (Budyko and Ronov 1979). Curve 2 in recalculated by Lapenis on the basis of Dobrodeev and Suyetova's data (1976) on the differences between the mean global temperature variations and temperature variations in the latitudinal zone covered with Emiliani's data. Curves 3 and 4 are obtained from the calculated carbon balance in the oceans derived from the studies of Lapenis (1984) and Berner et al. (1983). Although the accuracy of each of these four independent methods for evaluating carbon dioxide changes in the atmosphere is limited, together they permit the conclusion that CO_2 content has decreased by approximately five times over the last one hundred million years.

Since the four different methods yielded the same conclusions, it is possible to state that the principal trend of atmospheric carbon dioxide changes can be determined even by a very rough calculation technique. Carbon dioxide changes at the end of the Phanerozoic during shorter time intervals (on the order of ten million years) can be checked by the palaeoclimatic information given above.

Turning to the question of the atmospheric carbon dioxide content in the remote past, we mention that some time ago Sagan and Mullen (1972) and Sagan (1977) noted that it is difficult to account for the relatively warm Precambrian climate with greatly attenuated intensity of solar radiation. These scientists presented some data on Precambrian temperatures found by the isotope method, which show that the palaeotemperature then ranged from 280 to 350 K with a possible tendency to decreasing with time.

Since these data are not very accurate, they should be compared with the information available on the natural conditions of the Precambrian. It is known that since the Early Precambrian there had been liquid water and living organisms on the Earth and from time to time large glaciations partly covered the continents. Thus, in particular, it is possible that a considerable glaciation took place about 2.3 billion years ago. This glaciation probably consisted of a series of successive glacial epochs divided by time intervals of hundreds of millions of years.

It is quite probable that Precambrian primitive organisms could live in a rather narrow temperature range from 10° to 40°C. It might be thought that this range roughly corresponded to the range of the mean global temperature variations during the Precambrian after the appearance of organic life. This assumption is based on the conclusion drawn from modern climate theory that a complete glaciation of the Earth is possible with a relatively small decrease in the mean temperature of its surface (of approximately 6°-10°C compared with the present epoch). At the same time there is an upper limit of temperature increase, which is lower than the boiling point of water. At this temperature a decrease in solubility of carbon dioxide in sea water will lead to a considerable growth in atmospheric CO_2 concentration, at which the lower air layer will be heated to such an extent that water will boil in the warmest regions. An increase in the atmospheric water vapour volume will then intensify the greenhouse effect and lead to a further temperature increase. As mentioned in Chapter 1, under such conditions the climate of the Earth would be similar to that of Venus.

This upper limit of mean global temperature variations depends on a number of factors and is undoubtedly more than 40°C. However, the most ancient organisms could hardly exist under temperatures above 40°C, which are unfavourable for the majority of the existent primitive living beings.

According to Table 10, during the Phanerozoic the global surface air temperature averaged over geological epochs varied within a relatively narrow range equal to 11°C. With regard to the mean temperature of the coldest stages of Quaternary glaciations, this range increases to 16°C. The range of short-term fluctuations in the Precambrian mean global temperature was probably greater than that, but evidently did not exceed a few tens of degrees, because a greater temperature change would have led to the biosphere's destruction.

In the time interval of 4 billion to 1 billion years ago, solar radiation was 5% to 20% lower than at present. Other things being equal, this should have decreased the mean air temperature by 7° to 28°C compared to the present epoch, which corresponds to the absolute temperature range from 260 to 280 K. Since under such conditions stable global glaciation would have appeared, it is clear that there were no "equal conditions" in the Precambrian compared to the present time.

Because of the presence of large water bodies, the Earth's albedo in the Precambrian could not have differed much from the albedo at the beginning of the Phanerozoic (except for the glacial epochs, when the increasing albedo produced by ice area expansion was not a cause but a result of the initial climate cooling). Therefore, the most persuasive hypothesis of the causes of the warm Precambrian climate is that the greenhouse effect was much greater because of a perceptible difference in the atmospheric chemical composition in the Precambrian compared with the present time.

Such a supposition was made by Sagan and Mullen (1972), who thought that the Precambrian atmosphere contained ammonia, which intensified the greenhouse effect. Ammonia, however, is rather unstable and there could not have been any great amounts of ammonia in the presence of large quantities of atmospheric oxygen (which accumulated long before the end of the Precambrian). Therefore, this hypothesis was not acknowledged. It is more probable that in the remote past the atmosphere was much richer in carbon dioxide compared to the present time because of a higher level of volcanic activity. A similar opinion was expressed by Owen et al. (1979) and some other scientists.

As noticed by Walker (1983), in the Early Precambrian the cycling of atmospheric carbon dioxide could be similar in some respects to its present circulation, with CO_2 mass being 100-1000 times greater than at present.

If the assumption is made that during the Precambrian the air temperature averaged over long periods was invariable and differed little from the present one, it is possible to calculate the changes of atmospheric carbon dioxide with time.

Such an assumption is indispensable for explaining the continuous existence of living organisms throughout the Precambrian. As already mentioned, the temperature range for the existence of living organisms is rather narrow. Besides a general trend of atmospheric carbon dioxide variations because of gradual attenuation of degassing, the Precambrian undoubtedly witnessed rhythmical fluctuations in the degassing rate, which were accompanied by carbon dioxide changes and corresponding changes in the mean global temperature. Similar fluctuations of the degassing rate in the Phanerozoic were accompanied by mean temperature changes of approximately 16°C. In the Precambrian, under great variability of natural conditions, the atmospheric carbon dioxide fluctuations could have been much greater than at present, and the temperature changes probably approached the range favourable for the existence of living organisms. In this case life on Earth could have survived only with a more or less constant background temperature of the lower air layer, i.e. the temperature averaged over hundreds of millions of years.

In these calculations of temperature variations during the Phanerozoic we have used the following formula:

$$\Delta T = \Delta T' + \Delta T_s + \Delta T_\alpha, \tag{18}$$

where ΔT is the difference between the mean global temperatures at a given time interval and at present; $\Delta T'$, ΔT_s and ΔT_α are similar differences produced by changes in carbon dioxide concentration, the solar luminosity, and the Earth's albedo, respectively.

Considering the Precambrian albedo to be close to the albedo at the beginning of the Phanerozoic, let us assume that $\Delta T_\alpha = 4$°C. As already mentioned, the value of ΔT_s is equal to $-7t$°C, where t is the absolute age in billion years. The value of $\Delta T'$ is assumed to be $(3/\lg 2)\lg m_c$, where m_c is the ratio of CO_2 concentration during a given time interval to its concentration in the present atmosphere.

With this assumption, we find that

$$\frac{3}{\lg 2} \lg m_c + 4 = 7t. \tag{19}$$

From Eq. (19) we determine the dependence $m_c(t)$, which is shown as line 3 in Fig. 31. Although this dependence has been found on the basis of rather conditional assumptions, it is sufficiently realistic, because in order to compensate for the temperature effects of solar radiation changes, the range of CO_2 changes in the Early Precambrian should have been very wide. Therefore, the rough calculating technique cannot greatly affect the results obtained. Let us note that the calculated values of carbon dioxide concentration for the beginning of the given time interval are close to the magnitudes of Walker (1983).

For comparison, Fig. 31 presents the average mass of carbon dioxide during the Phanerozoic (line 1) and in the Late Proterozoic (line 2). It follows from this figure that until the Late Proterozoic the amount of atmospheric carbon dioxide was decreasing first rapidly and then more slowly. In the Late Proterozoic-Phanerozoic the CO_2 amount continued to decrease, but even more slowly. As already mentioned, throughout the Precambrian, besides a general tendency to decreasing CO_2 amount, rhythmical fluctuations took place, which were similar to CO_2 fluctuations in the Phanerozoic. Both the general trend of atmospheric CO_2 variations and its relatively short-term fluctuations were produced by changes in the degassing rate of the upper mantle, which were reflected in volcanic activity changes.

Let us note that although the data represented by line 3 are more hypothetical than those for the Late Proterozoic and Phanerozoic, they have been substantiated beyond the assumption of an invariable average climatic conditions of the Early Precambrian. For that time the dependence of CO_2 mass on time is assumed to correspond to an e-times decrease in volume in about 600 million years. Such a value does not differ very much from the characteristic time of changes in degassing rate used by Li (1972). This means that the dependence of the amount of CO_2 on time in the Early Precambrian, similarly to line 3, can be constructed without hypothesizing invariable climatic conditions.

We hope that more detailed estimates of atmospheric carbon dioxide variations during the Precambrian will be obtained on the basis of further investigations of the chemical composition of Precambrian sediments.

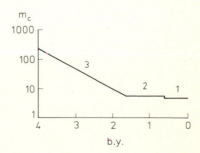

Fig. 31. Changes in relative mass of carbon dioxide

It is worth noting that the calculations by Eq. (19) allow certain conclusions about the accumulation of nitrogen in the Precambrian atmosphere. This formula describes temporal changes not only of the relative amount of atmospheric carbon dioxide but of the relative rate of CO_2 release into the atmosphere for the time interval of 1.5-4 billion years ago. The calculations based on this formula show that at the early stage of the history of the secondary atmosphere, the rate of degassing of the upper mantle was high but was rapidly decreasing with time. Such a change in the degassing rate was evidently associated with a decrease in the mass of radioactive elements in the Earth's crust. This rate was non-linearly dependent on the heat influx from radioactive disintegration and decreased faster than this heat flux attenuated. At the same time, in the Early Precambrian other heat sources occurred at the Earth's depths, the intensity of which was steadily dwindling.

If the composition of gases entering the atmosphere varied little throughout the history of the secondary atmosphere and the rate of atmospheric nitrogen loss was not great compared with the rate of its gain, it is possible to calculate by the data in Fig. 31 how fast the nitrogen atmosphere was forming.

Let us assume that the secondary atmosphere started to form about 4 billion years ago. Then we shall find that in 500 million years the nitrogen content of the atmosphere was slightly more than half its present content, in one billion years about 80%, in two billion years (i.e. two billion years ago), about 95%. These estimates show that the amount of atmospheric nitrogen varied much less than that of carbon dioxide and oxygen throughout most of the Earth's history.

3.2 Oxygen

Changes in Amount of Oxygen. As mentioned in Chapter 2, in the previous works of the authors of this book (Budyko 1977a; Budyko and Ronov 1979) the amount of atmospheric oxygen in the geological past was determined by Eq. (13) on the basis of data on organic carbon in continental sedimentary rocks. Since the absence of information on organic carbon in the oceanic sediments could have led to certain errors, the calculations were carried out in two versions. The first one suggested that the continental sediments comprise the major portion of the global organic carbon and the second assumed that the amount of organic carbon on the continents is a minor portion of the global mass and varies in proportion to this mass.

The calculated results are shown in Fig. 32. Curve M_O^I represents the results of the first version of calculation and curve M_O^{II}, the results of the second version. As can be seen from the figure, there is some difference between the curves, but they both reflect the same specific features of oxygen changes during the Phanerozoic. In these studies it was supposed that the actual change in oxygen amount corresponds to the value between M_O^I and M_O^{II}, but at that time it was very difficult to verify this supposition. Later, in order to find the oxygen amount, Budyko et al. (1985) used data on organic carbon for both the continents and the oceans.

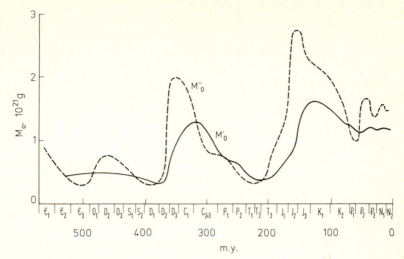

Fig. 32. Changes in oxygen mass in the Phanerozoic (M_o^I and M_o^{II})

Information on the organic carbon in oceanic sediments was, however, available only for the time interval from the Late Jurassic to the Pliocene. Since the earlier sedimentary rocks have not been preserved in the oceans, it was assumed on the basis of average data for the Late Jurassic-Pliocene that the total amount of oceanic sediments for all geological epochs from the beginning of the Phanerozoic to the Mid-Jurassic was equal to the total amount of continental sediments. The calculated results found by this method are presented in Fig. 33 (curve M_o^{III}).

When developing this study, a similar calculation was carried out on the basis of more accurate and detailed data on the amount of organic carbon in contin-

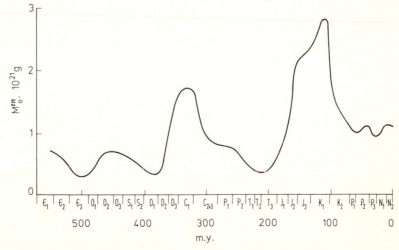

Fig. 33. Changes in oxygen mass in the Phanerozoic (M_o^{III})

ental and oceanic sediments given in Table 7. It proved that the organic carbon volume in the oceans for the Late Jurassic-Pliocene period is smaller than that on the continents and comprises approximately 45% of the global organic carbon mass. Therefore, for the geological epochs of the Palaeozoic and the first half of the Mesozoic, the global amount of organic carbon was considered to be 1.8 of the continental sediments. The parameter β_0 from Eq. (13) used in this calculation was found on condition that the oxygen mass at the end of the Cenozoic was close to its present mass. It proved to be 0.06 $(\text{m.y.})^{-1}$. When selecting the initial magnitude of M_{01} (oxygen mass for the beginning of the Cambrian), it has been assumed that this mass can be found for the Early Cambrian provided that $A_0 \gg dM_0/dt$. As mentioned in Chapter 2, the assumed magnitude of M_{01} for the beginning of the Phanerozoic hardly influences the results of calculating the oxygen amount for all geological epochs, except for the first one or two. The calculated results for different geological epochs designated by M_0^{IV} are included in Table 12 and represented as a curve in Fig. 34.

Table 12. Changes in amount of atmospheric oxygen

Stratigraphic interval		Duration of intervals (m.y.)	$\alpha_0 C_0$ 10^{20}g $(\text{m.y.})^{-1}$	$\gamma_0 M_c$ 10^{20}g $(\text{m.y.})^{-1}$	M_0^{IV} 10^{21}g	M_0^{V} 10^{21}g
E. Cambrian	ϵ_1	570−545 = 25	0.43	0.06	0.72	0.62
M. Cambrian	ϵ_2	545−520 = 25	0.29	0.09	0.63	0.51
L. Cambrian	ϵ_3	520−490 = 30	0.14	0.06	0.41	0.29
E. Ordovician	O_1	490−475 = 15	0.68	0.18	0.53	0.37
M. Ordovician	O_2	475−450 = 25	0.72	0.13	0.94	0.73
L. Ordovician	O_3	450−435 = 15	0.53	0.13	1.04	0.82
E. Silurian	S_1	435−415 = 20	0.24	0.09	0.77	0.57
L. Silurian	S_2	415−402 = 13	0.29	0.09	0.55	0.37
E. Devonian	D_1	402−378 = 24	0.14	0.09	0.41	0.25
M. Devonian	D_2	378−362 = 16	0.87	0.12	0.65	0.49
L. Devonian	D_3	362−346 = 16	1.06	0.18	1.24	1.02
E. Carboniferous	C_1	346−322 = 24	1.11	0.30	1.62	1.27
M.-L. Carboniferous	C_{2-3}	322−282 = 40	0.43	0.12	1.29	0.95
E. Permian	P_1	282−257 = 25	0.43	0.25	0.77	0.48
L. Permian	P_2	257−236 = 21	0.24	0.09	0.62	0.32
E. Triassic	T_1	236−221 = 15	0.24	0.09	0.47	0.27
M. Triassic	T_2	221−211 = 10	0.19	0.17	0.41	0.21
L. Triassic	T_3	211−186 = 25	0.43	0.11	0.51	0.30
E. Jurassic	J_1	186−168 = 18	0.72	0.12	0.83	0.63
M. Jurassic	J_2	168−153 = 15	1.59	0.16	1.50	1.28
L. Jurassic	J_3	153−133 = 20	1.07	0.23	1.91	1.62
E. Cretaceous	K_1	133−101 = 32	1.44	0.15	2.08	1.78
L. Cretaceous	K_2	101−67 = 34	0.59	0.18	1.74	1.46
Palaeocene	P_1	67−58 = 9	0.32	0.08	1.03	0.76
Eocene	P_2	58−37 = 21	0.59	0.12	0.93	0.71
Oligocene	P_3	37−25 = 12	0.53	0.03	0.94	0.77
Miocene	N_1	25−9 = 16	0.83	0.08	1.06	0.93
Pliocene	N_2	9−2 = 7	0.77	0.04	1.22	1.10

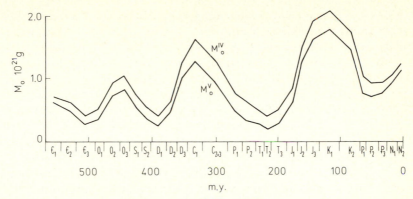

Fig. 34. Changes in oxygen mass in the Phanerozoic (M_0^{IV} and M_0^{V})

As has been shown in Chapter 2, at present we can apply the procedure for calculating the atmospheric oxygen mass, which accounts for the oxygen loss for oxidation of rocks as well as carbon monoxide and other not fully oxidized gases that enter the atmosphere from the Earth's crust. With this approach the calculation of oxygen amount can be made by Eq. (17).

Although such a technique for determining the oxygen mass is undoubtedly better than the earlier methods, its application is associated with certain difficulties. In particular, the available information on the ratio of the amount of not fully oxidized gases to carbon dioxide in the total mass of gases that are released from the Earth's crust is inadequate and does not permit a reliable estimation of the mean global value of the ratio. A more trustworthy method for determining this ratio is by data on the global oxygen balance. It follows from the data in Chapter 2 that oxygen consumption on oxidizing carbon monoxide and other not fully oxidized gases is from 0 to 43% of the total oxygen consumption on oxidation of all kinds of mineral matter. It is hardly probable that oxygen consumption on oxidizing the gases released from the Earth's crust approaches the upper or lower values of the indicated range.

Taking this into account, it is possible to rely on the hypothesis that the oxygen loss for oxidizing gases entering from the Earth's deep layers comprises about half of its greatest possible loss, i.e. 21.5% of the oxygen income to the atmosphere.

With this hypothesis the values of parameter γ_0 can be found by the following formula

$$\gamma_0 = 0.215 \ \frac{\alpha_0 C_0}{M_c} , \tag{20}$$

where $\alpha_0 C_0$ is the oxygen income rate averaged over the Phanerozoic equal to 0.60×10^{20}g (m.y.)$^{-1}$ and $M_c = 10 \times 10^{18}$g. Consequently, $\gamma_0 = 1.29$ (m.y.)$^{-1}$.

Since this estimate of parameter γ_0 is very rough, later we shall consider the question of how the error in determining this parameter affects the calculated results of atmospheric oxygen amount.

In the calculation of oxygen changes throughout the Phanerozoic by Eq. (17) it is possible to assume the value of M_{01} for the beginning of the Phanerozoic as in the previous case. The calculated results designated by M_0^V are given in Table 12 and represented as a curve in Fig. 34.

Let us consider the five estimates of oxygen changes in the Phanerozoic presented in Figs. 31-34. As already mentioned, these estimates were obtained by different methods and on the basis of organic carbon data, whose completeness was satisfactory to a different extent. It is important that all the calculations, without any exception, have revealed the same principal features of oxygen changes in the Phanerozoic. Among these features are the following:

(1) The tendency to an increasing amount of oxygen in the atmosphere prevailed throughout the Phanerozoic. At the same time, the growth of the oxygen mass was uneven and accompanied by repeated wave-like rises and drops in amount of oxygen.

(2) The lowest level in oxygen occurred in certain epochs of the Palaeozoic and at the Palaeozoic-Mesozoic boundary (the Late Permian-Triassic). The greatest rise in oxygen was observed in the second half of the Mesozoic, the second greatest oxygen rise occurred in the Early Carboniferous. The maximum amounts of oxygen were several times greater than its minima.

(3) During the Cenozoic the amount of atmospheric oxygen varied comparatively little and was greater than the average oxygen mass for the Phanerozoic.

By using data in Fig. 34, which reflect the most reliable calculated results of amount of oxygen, it is possible to somewhat extend these conclusions as to the principal features of atmospheric oxygen changes and present the following additional inferences.

(4) The oxygen amount varied throughout the Phanerozoic in the form of three waves, which gradually increased in altitude and length. The first wave, the weakest maximum in oxygen, took place in the Middle and Late Ordovician. The second, noticeably greater wave, occurred in the Early Carboniferous. The last wave, the greatest and longest rise, embraced most of the Jurassic and the entire Cretaceous periods with the oxygen maximum in the Early Cretaceous.

(5) The minima of oxygen amount in the Late Cambrian, Early Devonian and Middle Triassic were almost equal in value.

(6) It is possible that in the second half of the Cenozoic the oxygen volume began to increase, i.e. the oxygen minimum observed after the Mesozoic maximum was already overcome in the Palaeogene. If this conclusion is correct, the indicated minimum is the first, when the oxygen level appeared to be much higher than during the previous oxygen decreases.

In addition to these general conclusions, it is possible to decide which of the available quantitative data on changes in atmospheric oxygen during the Phanerozoic are most reliable. As mentioned above, the most complete data on the amount of organic carbon in the sediments are used in the last two calculations, whose results are presented in Fig. 34.

As to the methods for determining the oxygen amount, it should be noted that the first of them used when calculating the values of M_0^{IV} is a specific version of the second method (when $\gamma_0 = 0$, i.e. when the oxygen loss for oxidizing carbon monoxide and other not fully oxidized gases is insignificant compared with oxygen loss for oxidizing mineral rocks).

The second method used in the calculation of values M_0^V is based on the assumption that the indicated oxygen loss is approximately one fifth of the total oxygen loss in oxidizing mineral matter.

Since the calculated values of M_0^{IV} and M_0^V differ somewhat from each other, it should be indicated that the first method would have yielded rather accurate estimates if the relative mass of not fully oxidized gases was insignificant, for instance, if it was less than 1% of the carbon dioxide mass entering the atmosphere.

Since such a condition agrees badly with the available data on the composition of volcanic gases, it may be concluded that the values of M_0^{IV} are slightly overestimated, particularly for the epochs of a high level of atmospheric carbon dioxide concentration.

Let us note that the accuracy of calculating M_0^V decreases on determining the minimum values of M_0^V, when there appears to be a small difference between rather large values, each of which is found with considerable error.

Since the estimate of the relative amount of not fully oxidized gases adopted in the calculation of M_0^V is very rough, it is desirable to understand how the accuracy of estimating γ_0 affects the calculations of oxygen amount. For this purpose, it is possible to use Fig. 35, in which the values of M_0^{IV} and M_0^V are compared for 28 Phanerozoic epochs. As can be seen, these values are closely interconnected.

Thus it can be accepted that with γ_0 ranging from 0 to 1.3 (m.y.)$^{-1}$ the relative oxygen changes in the geological past depend little on the selected value of γ_0.

We may suppose that the actual mean value of γ_0 (or in other words, the ratio of the mass of not fully oxidized gases of amount of CO_2 that is released from the Earth's crust) for the entire Phanerozoic could not have been much greater than

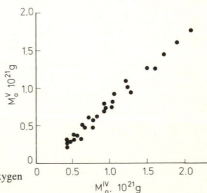

Fig. 35. Comparison of the calculated results of oxygen mass in the Phanerozoic (M_0^{IV} and M_0^V)

the value used in the last calculation of changes in oxygen mass. Otherwise, the oxygen amount in certain geological epochs (Early Devonian, Middle Triassic) would have been very small, which is incompatible with the presence of highly developed animal forms at that time. However the value of γ_0 could have been somewhat smaller than the values assumed in the calculation. Therefore, it is probable that the most reliable values of oxygen mass in the geological past either correspond to or are slightly greater than the data represented as curve M_0^V.

In this case the oxygen consumption on oxidation of mineral matter during the Phanerozoic is approximately made up of one-fifth by its consumption in oxidizing gases entering the atmosphere from the depths of the Earth's crust and of four-fifths by its consumption for oxidizing the mineral matter of the Earth's sedimentary layer. The study by Ronov (1982) (see Chap. 2) presents the estimates of the basic forms of oxygen expenditure in oxidation of rocks, which represents somewhat more than half of the total consumption. The problem of how the remaining, comparatively small, portion of the total oxygen income is consumed requires further investigation.

Let us now consider the causes which led to fluctuations in content of atmospheric oxygen during the Phanerozoic.

One of them is changes in photosynthetic productivity, which were mostly determined by the structure of the vegetation cover (its distribution and composition), by atmospheric carbon dioxide content and climatic conditions. The second important cause is changes in the global conditions of burial of organic carbon, which for most of the Phanerozoic depended greatly on the moisture conditions on the continents.

The influence of these factors is particularly pronounced during the time of the greatest oxygen fluctuations, which took place from the beginning of the Devonian to the end of the Mesozoic. It is evident that an increase in carbon dioxide content in the Devonian, which reached the highest Phanerozoic level in the Early Carboniferous, resulted in a considerable growth of global photosynthetic productivity, in particular because of the expansion of vegetation over the continents. The position of the continents and oceans made it possible for a humid climate to prevail in the Early Carboniferous over most of the land surface. This, on the one hand, enhanced the productivity of photosynthesis, and on the other, considerably increased the portion of organic carbon produced by photosynthesis and buried in sediments. This in turn gave rise to a high level of oxygen in the Early Carboniferous.

A still greater increase in atmospheric oxygen occurred in the second half of the Mesozoic. This was evidently associated with increased photosynthetic productivity (which, however, was generally explained by progressive evolution of plants and humid climate on the majority of the continents rather than by high CO_2 concentration) and particularly favourable conditions of organic carbon burial.

An abrupt decrease in oxygen mass during the Late Permian-Triassic was undoubtedly associated with the expansion of arid conditions on the continents. At the same time, this decreased photosynthetic productivity and conditions for accumulating organic carbon in sediments deteriorated. It is known, in particular, that the area covered by evaporites on the continents during the Triassic was much greater than during all other periods of the Phanerozoic (Crowley, 1983). A smaller decrease in oxygen content in the Cenozoic seems to be explained mainly by a relative reduction in the productivity of photosynthesis because of decreased carbon dioxide concentration in the atmosphere.

The question of oxygen content of the Precambrian atmosphere is more difficult to solve than that of changes in atmospheric oxygen content during the Phanerozoic.

Using the data of Ronov (1982), it is possible to evaluate the average mass of atmospheric oxygen for the Late Proterozoic, i.e. for the time interval from 0.57 billion years to 1.60 billion years ago. It follows from these data that the average rate of organic carbon accumulation in the Late Proterozoic sediments was 0.465 of the average rate of organic carbon accumulation during the Phanerozoic. At the same time this calculation shows that the atmospheric carbon dioxide content in the Late Proterozoic was 15% higher than in the Phanerozoic.

For period throughout the Late Proterozoic, the rate of changes in oxygen volume was much lower than its total gain or loss. Therefore, the oxygen content can be determined by the formula:

$$M_O = \frac{\alpha_O C_O - \gamma_O M_c}{\beta_O}. \tag{21}$$

Comparing the indicated values of $\alpha_O C_O$ and M_c with similar data for the Phanerozoic, we find from Eq. (21) that for the Late Proterozoic $M_O = 0.21 \times 10^{21}$g.

Using the data on variations in carbon dioxide concentration during the time up to the Late Proterozoic (see Fig. 31), it is possible to calculate when the atmosphere started to accumulate a noticeable amount of oxygen. In this calculation let us assume that during this time interval the oxygen income corresponded to the average Late Proterozoic income (i.e. it was 0.465 of the similar Phanerozoic value), and consequently equal to 0.279×10^{20}g (m.y.)$^{-1}$. Under such an assumption we find from Eqs. (19) and (21) that the indicated time interval is about 2 billion years. Taking this result into account, we can draw a schematic picture of atmospheric oxygen changes with time (Fig. 36).

Although the calculation of amount of oxygen during the Precambrian is rough, it reflects, in particular, the fact, known from many empirical studies, that the atmospheric oxygen content noticeably increased about 2 billion years ago. At the same time the general trend of oxygen changes in Fig. 36 differs greatly from that presented in earlier studies on the basis of hypothetical considerations.

For instance, it has not been confirmed that throughout most of the Earth's history the amount of oxygen in the atmosphere has increased continually. It seems that up to 2 billion years ago the atmospheric oxygen content was very low, because the oxygen produced through photosynthesis was entirely consumed in the course of oxidation of organic and mineral matter, including not fully oxidized gases that enter the atmosphere from the Earth's depths. The process of

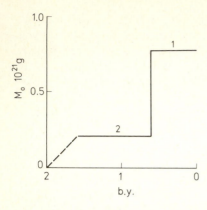

Fig. 36. Changes in oxygen mass M_o *1* Phanerozoic; *2* Late Proterozoic

rapid increase in atmospheric oxygen became possible only after volcanic activity decreased to the level when the entire mass of not fully oxidized gases received by the atmosphere could become oxidized by the available oxygen. In the Late Proterozoic atmospheric oxygen content reached a level constituting 15 to 20% of the present oxygen amount.

The process of increasing oxygen content that started about 2 billion years ago was certainly not continuous. As can be seen from the data given earlier, during the Phanerozoic the tendency to increasing oxygen content was disguised to a considerable extent by great but relatively short-term oxygen fluctuations. It is quite evident that similar fluctuations took place also throughout the Late Proterozoic, even though we have as yet no detailed data on the composition of sediments at that time, which are indispensable to disclose the indicated fluctuations.

Although the data on oxygen amount in the Precambrian (see Fig. 36) are very rough, they are sufficient to disprove the ideas presented in a number of studies that up to the end of the Precambrian the oxygen content represented a small portion (less than 1%) of its present extent.

A general conclusion from the above data is that, beginning from the time interval of about 2 billion years ago, the atmospheric oxygen amount tended to increase at an average rate, which changed relatively little over very long time intervals (hundreds of millions of years). During shorter time intervals the oxygen mass often considerably increased or decreased.

We should note that the estimate of the time of the appearance of oxygen atmosphere was obtained by rough calculation, therefore, it does not contradict the conclusions from some of the modern studies, which suggest that this time was somewhat greater than 2 billion years.

Oxygen, Carbon Dioxide and Life. As mentioned in Chapter 1, atmospheric oxygen was almost entirely created as a result of photosynthetic activity of autotrophic plants.

The amount of oxygen that enters the atmosphere daily depends on photosynthetic productivity, although this dependence is ambiguous, i.e. oxygen production to the atmosphere increases under conditions of prevailing humid climate on the continents (which accelerates the accumulation of organic carbon in the lithosphere) and decreases with the expansion of arid climatic conditions. At the same time the atmospheric oxygen volume is influenced by the level of volcanic activity. With intensification of vulcanicity, the atmosphere receives a greater amount of both carbon dioxide (which enhances photosynthesis and consequently increases the oxygen volume) and not fully oxidized gases, which, other things being equal, decreases the oxygen content of the atmosphere. Thus, the oxygen balance in the atmosphere is determined by a complex interaction between biotic and tectonic factors and climate.

Variations in atmospheric carbon dioxide concentration depend on both biotic and endogenetic factors, although in this case the role of abiotic factors is more important than their role in atmospheric oxygen balance. CO_2 production into the atmosphere is mainly determined by the degassing rate of the upper mantle and crust, i.e. by the processes occurring at considerable depths. At the same time the loss of atmospheric carbon dioxide depends on the process of weathering of rocks, on the productivity of autotrophic plants absorbing CO_2 and also on the activity of various organisms that accumulate organic carbon and carbon compounds, from which carbonate rocks are formed.

Taking into account that organisms influence oxygen and carbon dioxide balances, it is only natural to ask: What influence do atmospheric oxygen and carbon dioxide have on animals and plants? The idea of the influence of the atmospheric chemical composition on animate nature was already expressed in the first half of the 19th century by E. Geoffroy St.-Hilaire (1833), who wrote: "Let us suppose that in the course of a slow and gradual advancement of time the proportions of different components of the atmosphere changed, and it was an absolutely indispensable result that the animal world was affected by these changes". However, because of the absence of reliable information on changes in the chemical composition of the atmosphere in the geological past, the idea of Geoffroy St.-Hilaire for a long time attracted no attention. In recent years, the new data on oxygen and carbon dioxide changes during the Phanerozoic have been compared with the history of development of animate nature.

We present below some of the results of these investigations, some of which have been published earlier (Budyko 1980, 1982, 1984). It was suggested in these studies that with an increase in the atmospheric oxygen content, the possibility of developing aromorphoses increased, aromorphoses meaning crucial changes in organisms associated with enhanced energy level of their activity.

The idea of aromorphoses developed by Severtsev (1925, 1939) and Schmalgauzen (1940) was a great achievement of evolutionary biology. These scientists studied different examples of aromorphoses, paying particular attention to the appearance of the most progressive groups of animals, for instance, to the origin of the vertebrate classes.

If we follow the history of the development of organisms associated with the appearance of aromorphoses from the early stages, it becomes evident that the greatest step in this history was the dissemination of aerobic organisms. It is known that a great majority of modern organisms are aerobic, i.e. they use free oxygen in the reaction of decomposition of glucose and related substances. By this reaction CO_2 and H_2O are produced and energy is liberated that is used by organisms. Contrary to this, typical anaerobic organisms can live only in the absence of free oxygen. These include a number of microorganisms, which receive energy as a result of fermentation reactions based on disintegration of organic or inorganic substances, these reactions being less effective from the energy point of view. In addition to typical (obligatory) anaerobic organisms, there are various facultative anaerobic organisms, including multi-cellular ones, which can exist both with and without free oxygen in the environment, using it sometimes in the course of their life activity.

It is probable that many anaerobic organisms originated from the most ancient organisms that occurred during a part of the Precambrian, when the atmosphere contained only negligible amounts of free oxygen.

A wide distribution of aerobic organisms became possible only with the formation of oxygen atmosphere, i.e. about 2 billion years ago. Since the atmosphere accumulated considerable quantities of oxygen relatively quickly, it is possible that the development of ancient aerobic organisms was rather fast.

At the same time it is possible that facultative anaerobes could have appeared long before the oxygen atmosphere. It is suggested that such organisms lived in small "oases", i.e. the limited space adjacent to the plants, where the free oxygen released by photosynthesizing plants was not entirely absorbed in the course of oxidation of reducing gases.

Considering that a wide distribution of aerobic organisms was one of the most important aromorphoses, another important stage was the appearance and dissemination of multi-cellular forms. Sokolov (1975), Mayr (1976) and other authors expressed their assurance that the increase in oxygen mass was an essential factor for the appearance of multi-cellular organisms. It is probable that multi-cellular organisms appeared at the end of the Late Proterozoic and possibly at the beginning of the Vendian, i.e. about 680 million years ago.

It is easy to see that it is much more difficult for multi-cellular organisms to provide their internal organs and tissues with the required amount of oxygen than for uni-cellular organisms, which are usually small in size. As is known, the process of oxidation of any matter is accelerated by an increase in its surface/mass ratio, i.e. with its division into small portions. Therefore, the oxygen required for the process of oxidation of tissues of minute organisms can be obtained through the diffusion of this gas in the body of these organisms without any special adjustments necessary for large organisms. Since to develop such adjustments was a long process, it is natural to suppose that dissemination of relatively large and particularly of multi-cellular organisms could proceed faster with a pronounced increase in free oxygen in the environment.

As already mentioned, the Late Proterozoic undoubtedly witnessed considerable fluctuations in atmospheric oxygen, although until now we have no accurate data on the absolute age of the epochs with oxygen maxima. We can, however, advance a suggestion as to the time of the greatest increase in atmospheric oxygen in the Late Proterozoic. Let us emphasize that the grounds for this supposition are highly conditional and that this whole problem requires further investigation.

As can be seen in Fig. 34, throughout the Phanerozoic there were three successively increasing oxygen maxima. It follows from the data in the figure that at the beginning of the Cambrian, evidently, there was also a maximum of atmospheric oxygen, because the average amount of oxygen in the Early Cambrian was greater than its average amount for both the Late Proterozoic and Middle-Late Cambrian. Since the time interval between the subsequent oxygen maxima (the Late Ordovician and the Early Carboniferous) was about 110 million years, it is possible that the preceding maximum also occurred 110 million years earlier than the Ordovician maximum, which approximately corresponds to the beginning of the Cambrian.

Proceeding with such an extrapolation, one can suggest that the last oxygen maximum in the Late Proterozoic took place still 110 million years earlier, i.e. 680 million years ago. Since during the Phanerozoic the oxygen amount was gradually increasing from maximum to maximum (a natural consequence of the higher photosynthetic activity of plants), it is evident that the same principle can be revealed for fluctuations in atmospheric oxygen in the Late Proterozoic. In this case the oxygen maximum that took place 680 million years ago should have been the highest during the Late Proterozoic, although it was probably lower than all the Phanerozoic maxima, including the maximum at the boundary of the Vendian and Cambrian.

These considerations can be compared to another principle derived from the information for the Phanerozoic. It seems that each maximum of atmospheric oxygen was, with a certain lag, followed by glaciation. The greater the maximum, the greater was the glaciation. The intervals between oxygen maximum and the epoch of glaciation development were on the average some several tens of millions of years. For instance, after the oxygen maximum in the Early Cretaceous, glaciation started developing in the Palaeogene. The oxygen maximum in the Early Carboniferous was followed by a great glaciation at the end of the Carboniferous. There are also indications of glaciation which was developing at the end of the Ordovician and the beginning of the Silurian after the Ordovician maximum of oxygen. Less reliable data are available on the Cambrian glaciation.

From this point of view, it is worth noting the foregoing information on a considerable glaciation with an absolute age of about 615 million years, which could be associated with the oxygen maximum in the Late Proterozoic that occurred several tens of millions of years earlier.

The conjectural age of that maximum (680 million years) is close to the beginning of the Vendian, when, according to many scientists, the multi-cellular animals started disseminating.

We think that by making the available data on the organic carbon content in the Late Proterozoic sediments more detailed, this supposition can be verified by the same method that has been used for estimating the average amount of atmospheric oxygen in the Late Proterozoic.

The influence of changes in the chemical composition of the atmosphere on the evolutionary process in the Phanerozoic is easier to study than the problem of the effects of atmospheric factors on the origin of multi-cellular organisms. For this purpose we can use the above information on oxygen and carbon dioxide fluctuations in the atmosphere. This information has been applied to draw Fig. 37, which shows the values of M_o [the ratio of atmospheric oxygen amount calculated by Eq. (17) to the present oxygen mass] and M_c (the ratio of carbon dioxide concentration in the Phanerozoic to its present concentration).

As already mentioned, an increased oxygen content in the Early Cambrian evidently shows the occurrence of an oxygen maximum at the boundary of the Cambrian and the Vendian. It is natural to suppose that this maximum promoted a wide dissemination of skeletal animal forms in the Cambrian. This supposition is based on the fact that any considerable advancement in the morphophysiological structure of animals requires an additional energy influx in the course of their ontogenetic development. One of the simplest forms of providing a greater energy flux is an increase in oxygen content of the environment. If after achieving a useful adjustment for the kind of animal that requires additional energy consumption, the energy flux started to attenuate (in particular, at the expense of a decrease in oxygen amount in the environment), there are two possibilities for the future for such animals. If the development of a more complex structure of these animals does not result in new substantial advantages in the struggle for existence, they will die out with the deterioration of the abiotic conditions necessary for their survival. But if these advantages are sufficient to overcome the difficulties in sustaining the energy balance necessary for more advanced forms, these animals will continue to live even after noticeable deterioration in their environmental conditions.

In this connection it appears credible that the Cambrian saw the appearance of a great majority of the higher taxonomic groups of marine animals, including Archaeocyathids, Gastropods, Radiolaria, Sponges, Brachyapods, Trilobites, Ostracods, Coelenterates and so on. Some of these groups became extinct at the end of the Early Cambrian; for instance, almost all Archaeocyathids died out.

In the Ordovician the marine fauna was greatly enriched, which was possibly associated with an increase in atmospheric oxygen in the Middle and Late Ordovician. At that time almost all of the existent phyla and the majority of classes of marine invertebrates appeared. The first vertebrates, jawless fish-like animals, also disseminated in the Ordovician.

Turning to the problem of the influence on animal evolution of the last two maxima of atmospheric oxygen (which were the highest) during the Phanerozoic, we shall note that an important example of such an influence was the appearance of the principal classes of modern vertebrate animals.

As is known, almost all the existing species of vertebrates belong to six classes: cartilaginous fishes, bony fishes, amphibians, reptiles, mammals and birds. According to palaeontological records, these classes emerged within two relatively short time periods. The first of these embraces the Devonian and a part of the Carboniferous (two classes of fish, amphibians and reptiles). Then for a long time no new classes of vertebrate arose and only at the boundary of the Triassic and Jurassic did mammals appear, and in the Jurassic birds emerged. Thus, the formation of the vertebrate classes was completed.

It is natural to ask why the appearance of classes of vertebrate took place within these time intervals. It is possible to suppose that fluctuations in atmospheric oxygen content were to a great extent responsible for this.

As seen in Fig. 37, with a general tendency to an increase in atmospheric oxygen content in the Phanerozoic, the oxygen volume rose mainly within two intervals of time: in the Devonian-Early Carboniferous and at the end of the Triassic-Jurassic. These intervals are quite close to the time of the dissemination of vertebrates.

Such coincidence can hardly be accidental. An increase in the metabolism of animals was an important feature of their progressive development in the course of evolution. The phylogenetic development of the cardio-vascular system and other vertebrate organs, which ensures the higher energy consumption by more advanced forms, is well known. However, an increase in the use of energy could also have been achieved by an increase in the oxygen content of the atmosphere. This would have created some advantages in the struggle for existence for more complex organisms, whose vital functions, other things being equal, required more energy.

The specific study of changes in the metabolic level with the formation of new classes of vertebrates is hampered by the absence of direct data on the physiology of the earlier representatives of relevant classes. It might be supposed, however, that an increase in metabolism was indispensable for the transition of vertebrates from water to land and for the formation of endothermal animals. In other cases an increase in the oxygen amount might have improved the conditions of functioning of separate organs of more progressive animals without a considerable rise in the metabolic level of an organism as a whole.

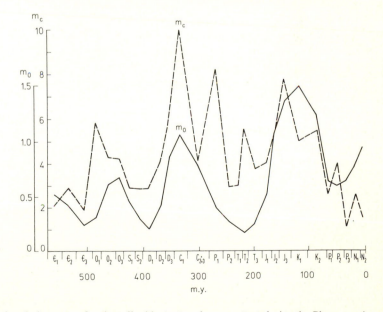

Fig. 37. Changes in relative mass of carbon dioxide (m_c) and oxygen (m_o) during the Phanerozoic

Therefore, it may be thought that an increase in atmospheric oxygen should have accelerated the evolution of animals and its decrease slowed it down.

Coming back to the question of the origin of modern classes of vertebrates, it might be suggested that the dissemination of earlier reptiles depended not only on the amount of oxygen in the atmosphere (although the oxygen content was close to maximum in the Mid-Carboniferous), but also on the first occurrence of arid conditions on part of the continents after the emergence of life on land. This supposition is in accordance with specific features of reptilian organisms that enable them to exist under conditions of insufficient moisture, as distinct from amphibians, which appeared earlier and were to a great extent aquatic animals in the Devonian and Carboniferous (Tatarinov 1972).

When treating the causes of the relationship between the epochs of the formation of aromorphosis of vertebrates and variations in atmospheric oxygen content, one might be reminded of the conclusion reached by Darwin (1859), that organisms are highly sensitive to even relatively small environmental fluctuations.

It is probable that of the different components of the external environment, the amount of oxygen in the air is one of the most essential factors of life activities for terrestrial animals (or the amount of oxygen in the water for aquatic animals, which depends on the atmospheric oxygen concentration). The level of metabolism of aerobic animals is directly dependent on the amount of oxygen in the environment, increasing, other things being equal, with an increase in oxygen content.

Additional energy received by an animal in the environment with increased oxygen content may be used for different purposes, including the development of a more complex structure of the organism or its individual organs and tissues in the course of evolution. Especially important in this case might be the improvement of organs maintaining metabolism, such as circulatory and respiratory systems in vertebrates.

According to Gilyarov (1975), in such cases the evolutionary process may take place as a positive feedback. As is known, these feedbacks intensify variations in the system produced by external factors, and as a result, the system may acquire a qualitatively new state. In the example under consideration an initial light improvement of circulatory and respiratory organs, induced by higher oxygen influx, additionally increases the energy flux to the organism, which creates conditions for further improvement and increased effectiveness of these organs. The influence of this positive feedback, in addition to the effects of gradually increasing oxygen content in the environment, was presumably the cause of the development of aromorphosis on the formation of new classes of vertebrates.

We shall not dwell on the description of the morphophysiological structure of aromorphosis, which is considered in detail in the works of Severtsev and Schmalgauzen already quoted. However, it should only be noted that the greatest changes in the energy level of life activity of vertebrates took place when amphibians emerged on land (where the presence of gravitational force sharply increased the amount of energy used on movement) and in the formation of endothermy in mammals and birds. It can be stated with certainty that these events could take place only with highly favourable changes in the environment, in particular with an increase in amount of oxygen.

It is no accident that the examples given here refer to vertebrates, as this is the most progressive group of animals, characterized by a high level of energy usage in the course of life activity. Therefore, the evolution of vertebrates seems to be dependent to a greater extent than that of invertebrates on factors limiting energy consumption. We think, however, that the evolution of invertebrates was also dependent on the external conditions affecting their energy regime.

Considering the influence of oxygen fluctuations on animate nature during the Phanerozoic, it may be supposed that in the epochs of increased oxygen content of the atmosphere, conditions appeared for enhancing the diversification of forms entering into the higher taxonomic groups, whereas in the epochs of decreased oxygen content, the diversity of forms grew less. This supposition is based on the idea that, with a large amount of oxygen, forms with relatively higher metabolic level might emerge, whose existence would have been impossible at a lower oxygen level.

Of considerable interest is the question about general principles of animal evolution throughout a long period from the end of the Carboniferous to the end of the Triassic, when the atmospheric oxygen level decreased for the longest time in the Phanerozoic. As a result, at the end of the Permian and throughout a part of the Triassic, the oxygen amount was only 20-25% of its present level, which is close to the average Late Proterozoic amount of oxygen. The lower (exothermal) vertebrates that used relatively little oxygen could more easily adapt to such conditions. It is very doubtful whether the higher (endothermal) vertebrates could exist at such low oxygen concentrations.

It is evident that with a decrease in atmospheric oxygen content, the life activities of animals and their mutual relations with other components of ecological systems must have been changing. A gradual deterioration of the conditions for the existence of animals must have been reflected in the specific features of the evolutionary process in the period in question.

These features were long noted by palaeontologists, although their causes were unknown. In particular, Simpson (1961) wrote: "It is interesting that no phylum has expanded steadily from the time of its appearance to the present day . . . The most nearly general feature is that most of the phyla contracted in the Permian, Triassic or both".

The principle indicated by Simpson was also confirmed by many other investigators. For instance, Raup and Stanley (1971) present a scheme of temporal variations in the number of taxons of fossil animals in the Phanerozoic, which shows that against a gradual increase in the number of taxons, their number abruptly declined in only one single period, in the Triassic. The authors indicate that, in fact, this contraction began in the middle of the Permian. Robinson (1971) states that from the Permian to the Late Triassic the number of genera of therapsids became ten times less and the number of genera of sauropsids increased considerably. This shows that conditions were unfavourable for the existence of mammal-like reptiles for the greatest part of the Triassic. Worth noting is also a tendency that appeared in the Triassic towards a decrease in size of the therapsids, which belonged to the most progressive groups. This tendency may be due to a growing shortage of oxygen, since the maintenance of the metabolism of large animals requires more oxygen than for small animals, other things being equal.

It might be supposed that a decrease in oxygen content during the Late Permian and the Triassic greatly retarded the process of the formation of mammals, which extended over an enormous interval of time, about one hundred million years. This class of vertebrate spread only with a new rise in atmospheric oxygen which began in the second half of the Triassic.

Another considerable decrease in oxygen concentration occurred during the second half of the Cretaceous. At this time the oxygen content decreased by more than 1.5 times and approached a level somewhat lower than the modern one. Although this level was rather high compared to the average conditions of the Phanerozoic, it is probable, however, that the oxygen decrease in the Cretaceous was of importance for the fauna of that time, and in particular caused the extinction of some groups of animals at the end of the Mesozoic.

Following these considerations, it might be concluded that the formation of modern classes of vertebrate was determined to a great extent by global changes in the environment, one of the causes of which was variations in the degassing of the Earth's mantle.

As mentioned earlier, with increases in the amount of oxygen in the atmosphere in the range of its variations throughout the Phanerozoic, possibilities emerged for the new groups of animals, which use more energy in the course of their vital activity, to evolve. Therefore, it is natural to suppose that in the epochs of increased oxygen concentrations, the diversity of fauna grew larger, while in epochs of decreased oxygen content their diversity diminished. To study this problem it is possible to make use of the data on animals that lived throughout the Phanerozoic, namely, marine invertebrates. The diagram in Fig. 38 has been constructed using information on the number of families of these animals for each period of the Phanerozoic (Pitrat 1970) and the data given earlier on variations in oxygen content. It shows the dependence of the number of families of marine invertebrates on the average amount of atmospheric oxygen in a given period

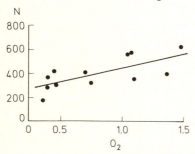

Fig. 38. The dependence of the number of families of marine invertebrates (N) on the relative oxygen concentration (O_2)

expressed in fractions of the present oxygen amount. The existence of a definite relation between the indicated values confirms the supposition made above.

Typical data on successive changes in diversity of certain groups of animals throughout the Phanerozoic are given in Fig. 39, which presents variations in the number of families of marine invertebrates, genera of echinoderms and orders of fishes according to the data from a monograph edited by Hallam (1977). As can be seen, the general principle indicated by Simpson is clearly pronounced for all these groups. The data in Fig. 39 show that the diversity of groups was greater in the epochs of increased oxygen content of the atmosphere. At the same time, the comparison of the data given in Figs. 37 and 39 makes it possible to conclude that there is a certain difference between variations in atmospheric oxygen and fluctuations in the diversity of animal groups in question. It can be observed that in addition to the effect of oxygen content, changes in diversity also exhibit a tendency towards a gradual increase in the number of taxonomic groups. As may be noticed, the number of taxons for the beginning of the Phanerozoic is smaller and at the end of the Phanerozoic is greater, compared to corresponding variations in atmospheric oxygen.

This tendency probably reflects in part the known fact of the steady loss of palaeontological information with time. Taking this into account, we may considerably improve the agreement between the data on oxygen variations and the diversity of the indicated animal groups.

The analysis of data concerning variations in the diversity of other higher taxonomic groups of animals shows that although "Simpson's rule" is fulfilled in most cases, we do not always observe a close correspondence between variations in the diversity of groups and fluctuations in atmospheric oxygen. The lack of such correspondence may be explained by the less strong influence of oxygen content on the evolution of organisms whose phylogenesis was less dependent on their metabolism (for instance, organisms that were small in size).

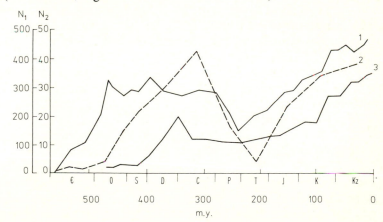

Fig. 39. Changes in the diversity of certain animal groups in the Phanerozoic (*1* families of marine invertebrates; *2* genera of echinoderms; *3* orders of fish; N_1 number of families and genera; N_2 number of orders)

It is worth noting that animals expending the maximum energy on movement disseminated during the epochs of the greatest increases in oxygen concentration.

These animals included, for example, flying animals such as large winged insects of the Carboniferous, pterosaurs and birds of the Jurassic, and also the largest land animals, such as the giant reptiles of the Jurassic and Cretaceous.

Turning to the problem of the influence of carbon dioxide changes in the atmosphere on animate nature, let us note that in some cases this influence was combined with that of atmospheric oxygen changes. Thus, in particular, in the foregoing example of the dissemination of aerobic organisms with the formation of oxygen atmosphere, the oxygen concentration in the atmosphere started to grow after a decrease in atmospheric CO_2 content below a certain limit, which resulted from a relevant decrease in the income rate of carbon monoxide and other not fully oxidized gases in the atmosphere.

During the Phanerozoic, carbon dioxide fluctuations considerably influenced photosynthetic productivity and climatic conditions. These fluctuations undoubtedly influenced the life of plants and, to a lesser extent, that of animals.

It can be seen from Fig. 37 that the first considerable maximum in atmospheric carbon dioxide concentration took place in the Ordovician. Such an increase in carbon dioxide concentration, besides enhancing the productivity of photosynthesis, produced climatic warming. This, other things being equal, should have created more favourable moisture conditions on the continents. All these factors were beneficial for the appearance of more advanced groups of plants, which could take over ecological niches previously inaccessible. Therefore, it is quite clear why it was in the Ordovician that the new forms of plants started to develop and disseminate throughout the continents.

Figure 37 shows that the most considerable increase in carbon dioxide concentration in the Phanerozoic began in the Devonian and reached its maximum in the Early Carboniferous. The second greatest rise in carbon dioxide was not so long and took place in the first half of the Permian. The third considerable maximum embraced the time interval from the Late Jurassic to the end of the Cretaceous.

In the Cenozoic, against a background of generally lowering CO_2 concentration, two relatively short maxima are pronounced in the Eocene and the Miocene.

The comparison of variations in carbon dioxide with principal events in the development of vegetation is impeded by the lack of a generally accepted chronology of the considerable changes in the nature of vegetation. Krasilov (1977) identified four major epochs in the formation of higher taxonomic groups of plants: the second half of the Devonian, the Permian to the beginning of the Triassic, the Cretaceous and the Miocene. The first of these epochs is associated with the formation of progymnospermous forests, the second with the expansion of coniferous forests, the third with the emergence of archaic flowering plants and the fourth with the appearance of steppe plant communities.

It may be noted that these epochs correspond to four out of the five maxima in carbon dioxide concentration that were enumerated. This testifies that an increased productivity of photosynthesis considerably affected the progressive development of plants.

At the same time the wide distribution of the flowering plants in the Cretaceous might have been promoted by a high level of atmospheric oxygen, which according to calculated data given in Fig. 37, took place in the first half of the Cretaceous. In this connection it should be noted that Takhtadzhyan (1980) considers that the angiosperms emerged at the beginning of the Cretaceous. Although a high level of oxygen concentration decreases photosynthetic productivity, it was essential for the development of flowers and germination of seeds of flowering plants, because plants absorb considerable amounts of oxygen when these processes take place.

The only maximum of carbon dioxide concentration that has no equivalent in the scheme of Krasilov is that of the Eocene. It is possible that this maximum was the factor promoting the formation of humid tropical forests in this epoch.

It is worth noting that our data confirm the idea of the relationship between the epochs of the appearance of higher taxons of plants and the tectonic processes. As mentioned earlier, fluctuations in volcanic activity corresponded to variations in carbon dioxide concentration, which in turn were in agreement with the development of vegetation cover.

Considering that the evolution of animals was to some extent associated with variations in atmospheric oxygen, we shall note that the dependence of fluctuations in atmospheric oxygen concentration on volcanic activity was of a complex and ambiguous nature. Although an increase in carbon dioxide concentration caused by intensive volcanic activity, other things being equal, might have raised the oxygen concentration due to an increase in photosynthesis, these "other things" were usually not equal. In particular, a release of photosynthetic oxygen into the atmosphere depended not only on the productivity of photosynthesis, but on the conditions of organic carbon burial as well, which sharply deteriorated when the climate became arid.

Although climatic conditions on the continents, and in particular the moisture conditions, depended on the topographic forms and the position of the continents, i.e. in the long run also on tectonic processes, this complicated relationship could not be expressed as a function of any one single variable.

In this connection it is hardly possible to find simple relations between the global characteristics of tectonic processes and the stages of the evolution of animals.

The conclusions we presented here on the relationship between the evolution of animals and plants and changes in the chemical composition of the atmosphere are only the first steps in understanding this problem. New achievements in studying the changes in atmospheric chemical composition will make it possible to investigate the relationship between the evolution of the atmosphere and that of organisms more thoroughly.

3.3 Past and Future of the Atmosphere

The Atmosphere and the Biosphere. The information already given allows us to estimate variations in oxygen and carbon dioxide concentrations in the geological past. As mentioned earlier, a regular decrease in atmospheric carbon dioxide was caused by gradual attenuation of the degassing rate of the Earth's deep layers associated with a reduced heating of these layers. At the same time, throughout the history of the secondary atmosphere carbonate sediments were accumulating at the expense of extracting carbon dioxide from the atmosphere.

Oxygen started amassing in the atmosphere, because the rate of gain of not fully oxidized gases lowered to the level at which practically all these gases could be oxidized by oxygen created by photosynthesis. Afterwards a further decrease in the entry of not fully oxidized gases into the atmosphere and the growth of photosynthetic productivity resulted in the accumulation of a considerable oxygen mass in the atmosphere.

Besides a general tendency to decreasing carbon dioxide and increasing oxygen volume in the atmosphere, the concentrations of these gases varied considerably over periods of 200 million years and much shorter time intervals.

Since the existence of all living organisms depends on the atmospheric chemical composition, the question naturally arises: why did the atmospheric chemical composition vary for about 4 billion years within a range permitting not only continuous preservation of life on the Earth (i.e. the continuous existence of the biosphere), but the progressive development of organisms, among which many achieved a high level of complexity in the course of a long evolution?

Although this question has rarely been discussed in scientific studies, it could be answered in two different ways. One of them is the so-called "hypothesis of Gaia", which suggests that living organisms have the ability of controlling the environment, and of maintaining a state favourable for their life activity (Margulis and Lovelock 1974; Lovelock 1979). The authors of this hypothesis have not adduced any definite proofs in favour of their idea, and it is mainly based on the apparent impossibility of otherwise accounting for the long existence of the biosphere.

Our conclusions about the mechanism of the evolution of the atmosphere have indicated that variations in the physical state and chemical composition of the atmosphere depend mostly on two external factors, namely, the evolution of the Sun, which leads to a gradual increase in solar radiation, and the evolution of the Earth, in the course of which the process of degassing of the upper mantle gradually attenuates. The first of these processes is entirely independent of the activity of terrestrial organisms and the second almost so.

The causes of the antiquity of the Earth's biosphere can therefore be explained otherwise. It might be thought that this antiquity is a result of random coincidence (independent of the existence of organisms) of the direction and rate of the processes of the Sun and the Earth's evolution, which are not connected to each other. Since the probability of such a coincidence is extremely low, it means that life (and particularly its higher forms) in the Universe is an exceptionally rare phenomenon.

This point of view has been developed in the works of Budyko (1977a, 1980, 1984) and Hart (1978, 1979).

As has been noted in these works, the atmosphere at which life on any planet can exist must have a specific physical state and chemical composition. As already mentioned, a decrease in the Earth's mean temperature by several degrees can lead to complete glaciation. An increase in this temperature by a few tens of degrees would lead at the present time to the destruction of organisms by overheating and in the remote past, when the atmosphere's mass was not great, the loss of liquid water as a result of its boiling.

It has been noted in the previous chapter that because of the increase in solar radiation the Earth's mean temperature, other things being equal, should rise by approximately 7°C per billion years, i.e. by about 30°C over the time of the biosphere's existence. It is easy to see that if the chemical composition of the ancient atmosphere was identical with its present composition, the Earth would have been covered with dense ice beginning from the remote past. Calculations based on the climate theory have shown that since the global ice cover decreases the mean temperature near the Earth's surface by several tens of degrees, the "white Earth" would also survive with the present level of solar radiation and only in a few billions of years could a further increase in solar radiation cause the ice to melt. In this case an increase in the Earth's surface temperature might be so great that the appearance of life on Earth would be scarcely possible.

Let us conclude that life could originate on Earth and be preserved for billions of years because of the coincidence of several factors. It might well be considered that the probability of each of these coincidences was very low.

The first of these coincidences is that about 4 billion years ago the greenhouse effect of the atmosphere, produced by a great mass of carbon dioxide and other gases absorbing the longwave radiation, increased the Earth's surface air temperature to such a level that this temperature, under conditions of reduced solar radiation as compared to the present epoch, proved to be in the range favourable for the existence of primitive organisms (possibly from 10° to 40°C). The slightness of the probability of such a coincidence is explained by the fact that the indicated temperature range is very narrow compared to the range of temperature changes on celestial bodies.

The second, still less probable, coincidence is that for 4 billion years the effect of the diminishing rate of degassing of the Earth's upper mantle on the Earth's surface temperature was approximately equal in value and opposite in sign to the effect of increasing solar radiation on the indicated temperature.

It might be thought that the presence of such a compensation, which has maintained a more or less unvarying climate on the Earth over billions of years, is an exceptionally rare event, whose reccurrence on any other planet is almost impossible.

The third favourable condition for the preservation of life is associated with the fact that the range of variations in atmospheric chemical composition was limited and did not exceed a rather narrow range that permitted the existence of organisms. For example, a relatively small decrease in atmospheric carbon dioxide mass compared with its present level (particularly with the lower level in the glacial epochs) would have led to the destruction of autotrophic plants, which would have ruined almost all organisms on the Earth.

The fourth condition for the preservation of life is that during the time of the biosphere's existence the scale of short-term fluctuations in the chemical composition of the atmosphere and its physical state (i.e. climate) was limited. Such fluctuations, as mentioned in Chapter 1, undoubtedly took place in the past and repeatedly led to the extinction of many organisms.

The fifth condition, in contrast to the previous four, was indispensable, not for the biosphere's preservation but for its development. As was shown in the previous section, the most important stages in the progressive development of organisms were associated with favourable changes in the environmental conditions. For instance, the formation of an oxygen atmosphere made it possible for aerobic organisms to disseminate. In other cases, an increased content of atmospheric oxygen and carbon dioxide was an essential factor in the progressive development of animals and plants. We may state that unvarying environmental conditions beginning with the appearance of the first living beings would have made it impossible for any complex organisms to evolve over the time of the biosphere's existence.

All these considerations show that the Earth's biosphere is unique and that the existence of other biospheres in our galaxy and in the system of nearest galaxies is hardly probable. A similar conclusion has been reached by astrophysicists (Hart 1975; Shklovsky 1976 and others), who were convinced that if extraterrestrial civilizations existed, we should have perceived certain evidence of their activity.

It should be elucidated that these considerations do not deny the possibility of certain relationships sustaining the Earth's biosphere's stability, including the relationships that are dependent to some extent on the activity of living organisms. Without such relationships, even the relatively short-term existence of the biosphere would be impossible, to say nothing of its long-term existence.

Thus, in particular, a relatively stable content of atmospheric carbon dioxide and oxygen is maintained by negative feedback between the amounts of these gases and their expenditure (see Chap. 2). However, even in the presence of these feedbacks the atmospheric concentrations of both carbon dioxide and oxygen varied by approximately ten times during the Phanerozoic, which influenced living organisms considerably and not always favourably.

For instance, the diversity of the majority of phyla of animals was reduced at the end of the Palaeozoic and the beginning of the Mesozoic, when the atmospheric oxygen content considerably decreased. In addition, a high level of atmospheric oxygen created by autotrophic plants, which occurred throughout a considerable part of the Phanerozoic, lowered photosynthetic productivity. The photosynthetic rate was also decreased during the epoch of diminishing atmospheric carbon dioxide content. This was induced, in particular, by the extraction of carbon dioxide from the atmosphere by the same autotrophic plants in the course of their activity. This example is interesting because it shows that plants are not able to maintain the chemical composition of the atmosphere which is favourable for their life. It can be easily understood why such an ability is absent in organisms. In the course of the struggle for existence, the organisms adapt to the current environmental conditions and they cannot foresee the remote consequences of interactions between living nature and abiotic medium, which can be harmful for them. It should be recalled that not only animals and plants, but man as well has only recently possessed such an ability. This has been the cause of repeated ecological crises, including such as were disastrous for certain nations in some historical epochs.

In the past, man's economic activity resulted only in local changes in the biosphere, although these local changes gradually expanded over increasing continental and oceanic areas.

In the 20th century, particularly in its second half, the consequences of industrialization embrace the entire Earth. These consequences appear to be associated with changes in the atmospheric chemical composition produced by the combustion of increasing amounts of coal, oil and other kinds of carbon fuel.

The Present Changes in the Chemical Composition of the Atmosphere. The conclusion that a pronounced change in the atmospheric chemical composition is inevitable as a result of the combustion of carbon fuel was first made by Arrhenius (1908) at the beginning of this century. Arrhenius considered that the ocean does not absorb all of the carbon dioxide produced in industrial processes. As a result, carbon dioxide concentration increases in the atmosphere. Arrhenius suggested that in this connection the global climate becomes warmer and photosynthetic productivity increases. Later Callendar (1938) expressed the opinion that climate will tend to warm because of increasing atmospheric carbon dioxide content produced by the combustion of carbon fuels. However, the works of Arrhenius and Callendar, being ahead of their contemporaries, did not attract the general attention to the problem of anthropogenic change in the atmospheric chemical composition.

The question of anthropogenic increase in atmospheric carbon dioxide content appeared clearly only in the 1960's, when first information was accumulating as a result of systematic observations of atmospheric carbon dioxide concentration, undertaken at the end of the 1950's within the framework of the International Geophysical Year. This information revealed a regular annual increase in carbon dioxide volume at a rate practically identical at stations situated at great distances from each other (Hawaii, Alaska, the South Pole and so on).

At the beginning of the 1970's, an increase in carbon dioxide concentration and a rise of the mean surface air temperature were estimated as being expected in the next century (Budyko 1972). The calculated results are presented in Fig. 40, where curve 1 designates secular variations in the mean air temperature anomaly

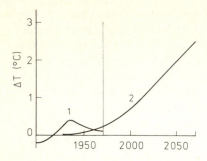

Fig. 40. Changes in mean air temperature ΔT (*1* mean air temperature anomaly in the Northern Hemisphere based on observational data; *2* expected temperature change produced by anthropogenic carbon dioxide concentration increase)

in the Northern Hemisphere based on observational data and curve 2 shows the expected temperature changes caused by anthropogenic growth of carbon dioxide concentration. As can be seen from this figure, the conclusion was made that the air temperature will increase by 2.5°C in 100 years. This corresponds to an approximate doubling of the CO_2 concentration.

During the second half of the 1970's, there appeared many other calculations of the impending changes in CO_2 concentration, carried out by both individual scientists and scientific bodies working on national and international commissions. Reviews of works on the problem of anthropogenic carbon dioxide increase are to be found in a number of monographs (Energy and Climate 1977; Budyko 1980; Changing Climate 1983 and others).

We give here the conclusions of recent Soviet-American specialist meetings on climatic effects of anthropogenic changes in the chemical composition of the atmosphere (Climatic Effects of Increased Atmospheric Carbon Dioxide 1982; Anthropogenic Climate Change 1984).

It is noted in the first of these reports that over the period of systematic observations of atmospheric carbon dioxide (1958-1980), its concentration has increased by 22 ppm, reaching a level of about 340 ppm. Industrial effluents of carbon dioxide for this time constituted 39 ppm of the atmospheric volume. Studies of the carbon cycle have shown that the remaining portion of anthropogenic carbon dioxide (17 ppm) is absorbed by the ocean, the role of the biota in the organic carbon balance being insignificant. It has been concluded on the basis of analyzing observational data of the 19th century and calculations of carbon dioxide balance, that in the pre-industrial time the carbon dioxide concentration was 290 ppm. Consequently, up to 1980 it has increased by approximately 50 ppm.

We shall note that in subsequent works the value of pre-industrial CO_2 concentration is considered to be slightly less than the estimate given here. This corresponds to a more considerable increase in atmospheric carbon dioxide concentration that took place over the last 100 years.

In these reports, the conclusion has been made that from 50 to 60% of the ejected anthropogenic carbon dioxide remains in the atmosphere; by the end of the 20th century the atmospheric CO_2 content will be 380-400 ppm, and by the middle of the 21st century carbon dioxide concentration might be double that of the end of the 19th century.

It is mentioned in the same reports that, according to a number of recent studies, man's economic activity is now leading not only to an increase in carbon dioxide volume but to an increase in some other atmospheric components, including methane, nitrous oxide, chlorofluorohydrocarbons (often called freons) and so on. All these gases are contained in the atmosphere in very small amounts. However, since they actively absorb longwave radiation in the range of 8-13 μm, even a moderate increase in the amount of these gases will lead to a noticeable intensification of the greenhouse effect, which will additionally raise the temperature of the lower air layer.

The first of these reports presents estimates of the expected increase in the mean surface air temperature compared to the temperature at the end of the 19th century. These estimates are based on the most probable prognosis of imminent change in carbon dioxide concentration and the indicated trace gases. In this calculation climatic effects of the thermal inertia of the Earth-atmosphere system have been taken into account. As a result of this calculation, the following values of the mean temperature change have been obtained:

	Year	2000	2025	2050
ΔT	°C	1-2	2-3	3-5

As already mentioned, in recent years many forecasts have been published of the change in mean air temperature in the next century, and these forecasts agree well with each other.

It follows from these studies that in the course of the next decades the effects of man's economic activity on the chemical composition of the atmosphere will produce climatic changes as great as those earlier occurred over many thousands or millions of years. Since such changes in climate will affect long-term economic planning, it is evident that a knowledge of future climate changes is of great practical value.

It is apparent that the available estimates of mean global air temperature changes are insufficient to answer the acute practical problems concerned with the future climate. For this purpose information is necessary on the expected changes in a number of climatic elements, and this information should cover definite geographical regions. The task of obtaining such information appears to be rather difficult. The basic methods for its solution are based on the usage of detailed climatic models and empirical analysis of observational data on climatic changes that occured over the last 100 years. The accuracy of the results produced by these methods is inadequate and must be checked by other independent methods.

A new independent approach to the solution of the problem of estimating future climatic conditions might be based on studying the natural conditions of the remote past, when the high carbon dioxide content of the atmosphere made climatic conditions much warmer than at present.

This method of studying future climatic changes has been used when constructing the maps of the expected changes in air temperature, precipitation and evaporation over the greatest part of the territory of the USSR (Budyko et al. 1978) and also in some other studies carried out in recent years.

In substantiating this method, information on the chemical composition of the atmosphere and climatic conditions of the Neogene is of great importance, i.e. for the time interval of 2-25 million years ago. As can be seen from the data given in Table 10, at that time the mean global temperature was 3° to 6°C higher than at present. Such a temperature increase corresponds to the warming which will occur in the 21st century. The climatic maps for the Neogene can thus be used to forecast climatic conditions for the next century.

The task of constructing palaeoclimatic maps presents a number of difficulties. The problem of the correct application of palaeoclimatic information in the calculations pertaining to future climate is also rather complicated. Without dwelling on the details of relevant investigations, we shall note that they usually reach a conclusion about the future climate which agrees with the inferences made on the basis of the other independent methods mentioned earlier for estimating climatic conditions of the next century.

Let us give some examples of the application of palaeoclimatic data in studying present climate changes, as presented in one of the Soviet-American reports (Climatic Effects of Increased Atmospheric Carbon Dioxide 1982).

Table 13 shows estimates of changes in the mean latitudinal surface air temperature and Table 14 gives changes in the mean latitudinal precipitation on the continents with a 3°C increase in the mean global temperature. Such a temperature increase corresponds to doubling the atmospheric carbon dioxide concentration compared with its pre-industrial level.

The results calculated by climatic models included in these tables were derived from Manabe and Wetherald (1980). The palaeoclimatic data used in

Table 13. Changes in mean latitudinal surface air temperature (°C) with mean global temperature increase by 3°C

	Northern latitude								
	0-10	10-20	20-30	30-40	40-50	50-60	60-70	70-80	80-90
Calculation by climatic model	1.7	2.0	2.5	3.1	3.8	4.3	5.2	6.8	7.6
Palaeoclimatic data	1.2	1.4	1.6	2.4	4.0	4.6	6.8	7.9	9.5

Table 14. Changes in mean latitudinal precipitation (cm year^{-1}) on the continents with mean global temperature increase by 3°C

	Northern latitude						
	10-20	20-30	30-40	40-50	50-60	60-70	70-80
Calculation by climatic model	10	14	3	−1	10	9	13
Palaeoclimatic data	12	12	2	2	8	9	13

these calculations have been obtained by Borzenkova and Yefimova from palaeogeographic information for the Pliocene. As can be seen, both of the methods for estimating climatic conditions of the future have yielded similar results for changes in mean latitudinal temperature distribution. It follows from both calculations that total precipitation on the continents increases mainly in low and high latitudes. In the latitudinal belt of 30°-50°N during warming, precipitation changes comparatively little and possibly even decreases slightly.

It has repeatedly been noted that the warming caused by man's economic activity which is occurring at present and will be much more intensive in coming decades, will correspond to the restoration of climatic conditions of the remote past. Throughout the greatest part of the next century, these conditions will be similar to the Neogene climate and subsequently will possibly to a certain extent resemble the still warmer climate of the Eocene.

The main cause of the present warming is the restoration of the ancient chemical composition of the atmosphere, when it contained a much greater amount of carbon dioxide than at present. An increase in atmospheric carbon dioxide is accompanied not only by climatic warming, but by increasing precipitation in the majority of latitudinal zones. Besides affecting the climate, an increase in carbon dioxide volume enhances the productivity of autotrophic plants, and this might lead to augmenting the global biomass of plants and animals.

It is worth mentioning that the rate of change in chemical composition of the atmosphere occurring at present is unusual, seen historically. Nowadays, within only a few decades the amount of atmospheric carbon dioxide is increasing by a value corresponding to its decrease previously over millions of years. The cause of this fast rate of modern evolution of the atmosphere is simple: by means of his economic activity, man is returning to the atmosphere the amount of carbon dioxide that has been extracted from it by photosynthetic plants for many millions of years. As already mentioned, man's impact on the atmospheric chemical composition is leading to the same rapid climatic warming over the entire Earth and to the same great changes in the biosphere, which in some respects resembles the biosphere of the remote past.

Conclusion

Throughout the greatest part of its history, the Earth has had a secondary atmosphere, created as a result of interaction between the process of degassing and the processes occurring at the Earth's surface. In the course of these processes new components (in particular, oxygen) appeared in the atmosphere, and all the atmospheric constituents, including the gases entering the atmosphere from the Earth's depths, have been consumed and returned to the lithosphere.

The total volume of the secondary atmosphere and its chemical composition were mainly determined by the circulation of atmospheric gases on the Earth itself, and depended little on the exchange of matter between the Earth and outer space.

The cycling of gases comprising the secondary atmosphere was considerably influenced by living organisms, the existence of which in turn depended essentially on the chemical composition and physical state of the atmosphere.

The basic atmospheric components, the amount of which depends on the vital activity of organisms and in the presence of which this activity is possible, are carbon dioxide and oxygen.

This book has presented the results of investigations on the basis of which it is possible to obtain quantitative estimates of changes in carbon dioxide and oxygen mass throughout the Earth's history. These estimates gave rise to the following conclusions:

(1) The composition of the secondary atmosphere varied greatly with time both in the Precambrian and in the Phanerozoic. The opinion that throughout a considerable part of the Earth's history, and in particular during the Phanerozoic, the atmosphere practically did not change appears to be erroneous and typical of the incorrect approach to studying the Earth's history from a uniformitarian stand point.

Having substantiated the known idea that the history of the secondary atmosphere is divided into two major stages: the oxygen-free atmosphere and the oxygen atmosphere, it was established, first, that the transition from the first to the second stage about 2 billion years ago proceeded rather quickly, and second, long before the beginning of the Phanerozoic, probably before the beginning of the Late Proterozoic, the atmosphere already contained a considerable amount of oxygen.

(2) A typical feature of oxygen-free atmosphere was first rapid and then slower decrease in amount of carbon dioxide. During the Precambrian the amount of atmospheric carbon dioxide decreased by approximately 100 times.

The decrease in carbon dioxide content was accompanied by a gradual accumulation of nitrogen in the atmosphere. Therefore, at the end of the epoch of oxygen-free atmosphere, nitrogen was the atmosphere's principal component. It could be supposed that throughout the period of oxygen-free atmosphere, fluctuations in the level of volcanic activity were accompanied by considerable relatively short-term variations in the amounts of atmospheric carbon dioxide and other gases released from the Earth's depths.

(3) The conditions for the appearance of the oxygen atmosphere were created long before the end of the epoch of oxygen-free atmosphere as a result of the emergence and wide distribution of autotrophic plants.

From the beginning of the epoch of oxygen atmosphere, the earlier existing not fully oxidized atmospheric gases, in particular carbon monoxide, practically vanished. The process of increasing the amount of atmospheric oxygen was slower in the Late Proterozoic and the Early Palaeozoic. A considerable increase in oxygen mass occurred after the appearance of a vegetation cover on the continents, which greatly increased global photosynthetic productivity. The amount of oxygen averaged over geological epochs (as well as the amount of carbon dioxide in the oxygen atmosphere) varied within a wide range because of the variability of volcanic activity and other factors.

(4) Throughout the greatest part of the history of the secondary atmosphere, its chemical composition was considerably influenced by living organisms. If an explanation of the appearance of life on Earth is difficult, it is even more difficult to account for the fact of preservation of life for about 4 billion years under conditions of constantly varying chemical composition and physical state of the atmosphere.

This can probably be explained by the following facts. Living organisms can, to a certain extent, adapt to fairly gradual environmental changes. However, the range of such variations is rather limited, which is seen in particular in the fact that even after billions of years of evolution the organisms have not been able to adapt themselves to live constantly in certain areas of the Earth's surface (the central part of the Antarctic and some other places). It is certain that the life zone (the range of changes in the physical and chemical conditions within which the existence of organisms is possible) comprises a very small part of the range of possible changes in the physical and chemical conditions on a planet of terrestrial type produced by the evolution of its atmosphere.

It follows from the foregoing calculations that the preservation of a comparatively constant climate on the Earth and changes in the chemical composition of the atmosphere, within a range permitting not only the existence of organisms but promoting their progressive development, are the result of an almost improbable coincidence of independent factors of the evolution of the atmosphere. The occurrence of such a coincidence is evidently explained by the vast number of planets in the stellar systems of the Universe, which makes it possible for the almost improbable events to occur on some of them.

In this connection it might be supposed that the atmosphere (as well as the biosphere) of the Earth is possibly unique in the galaxy or in our system of galaxies.

(5) The future inevitable disappearance of the atmosphere (and the biosphere, which is associated with the atmosphere) will be the result of the coming extinction of the Sun's radiation, after which all the atmospheric gases will solidify. However, it follows from the preceding conclusion that long before this event occurs (which will happen in a very distant future), changes in the atmosphere can arise which will lead to the destruction of all organisms.

To clarify this conclusion, we note that on the basis of the data presented in Fig. 37, two different suppositions can be made about the possible future of the atmosphere. Both of the suppositions are based on the extrapolation of the trends of carbon dioxide and oxygen variations during the Phanerozoic.

Firstly, it might be assumed that at present the degassing rate has reached its minimum and its future rise will cause an increase in carbon dioxide concentration. The resultant growth of photosynthetic productivity might accelerate the increase in atmospheric oxygen mass, which started in the Eocene and in millions of years will reach a maximum exceeding that of the end of the Mesozoic. In the natural course of the biosphere's development, this would lead to the emergence of new plants and animals, which would be in a further stage of development than the present ones.

The second supposition is based on the assumption that the present trend of lowering the degassing rate still continues, and atmospheric CO_2 concentration will go on dropping. In this case, glaciation might occur and spread over the entire surface of the Earth, after which the temperature at all latitudes will reach several tens of degrees below zero. Therefore, life on Earth will be destroyed and, quite probably, never appear again. Within the natural evolution of the atmosphere, such a prospect might come true in the geologically brief time interval of millions or even hundreds of thousands of years.

For the more distant future, the reality of such a possibility is greatly increased, because the resources of long-lived radioactive elements in the Earth's depths are gradually dwindling. This might lead to a disruption of degassing and, hence, to the destruction of the biosphere due to a decrease in the atmospheric CO_2 volume to below the level at which photosynthesis can take place or to below the level at which the advancement of global glaciation is precluded.

It is thus probable that the biosphere will cease to exist as a result of changes in the chemical composition of the atmosphere long before the extinction of solar radiation.

(6) The problem of the atmosphere's future has acquired new meaning during this epoch of the increasing impact of man's economic activity on the environment.

The rapidly intensifying effect of the combustion of carbon fuel on the atmospheric chemical composition demonstrates that natural conditions can well be altered over the entire Earth with the help of accessible technological means. It is certain that in the near future the further advance of scientific and technological progress will create a much greater possibility of influencing the chemical composition and physical state of the atmosphere.

Inadvertent changes in atmospheric composition associated with an increase in carbon dioxide concentration might give rise to certain favourable effects (an enhancement of photosynthetic productivity, warming in the countries with a cold climate, elimination of a possible glaciation development). Still greater success might be achieved in the future by working out efficient methods to influence the atmosphere for the sake of the whole of mankind. For the immediate future the realization of such a possibility will eliminate the danger of the destruction of the biosphere as a result of the natural evolution of the atmosphere.

Realizing the possibility of controlling the chemical composition of the atmosphere and its physical state will be a great step towards the formation of the noosphere; and this in turn will realize the idea of the founder of biosphere science, V.I. Vernadsky.

References

Amosov GA, Melekhova KD, Dobryakova NY (1980) Clarks of organic carbon and bitumen in sediments. Izv Akad Nauk SSSR Ser Geol 7:120 (R)

Anatolyeva AI (1972) Pre-Mesozoic red beds. Nauka, Novosibirsk, 348 p (R)

Anthropogenic Climate Change (1984) In: Budyko MI, Gates VL et al. (eds) Meteorol Gidrol 6:117–123 (R)

Arrhenius S (1896) On the influence of the carbolic acid in the air upon the temperature of ground. Philos Mag 41:237–275

Arrhenius S (1903) Lehrbuch der kosmischen Physik. Hirzel, Leipzig, 1026 p

Arrhenius S (1908) Das Werden der Welten. Hirzel, Leipzig, 108 p

Augustsson T, Ramanathan V (1977) A radiative-convective model study of the CO_2-climate problem. J Atmos Sci 34:448–451

Axelrod DI, Baily HP (1969) Paleotemperature analysis of Tertiary floras. Palaeogeogr Palaeoclimatol Palaeoecol 6:163–195

Berkner LV, Marshall LC (1965a) On the origin and rise of oxygen concentration in the Earth's atmosphere. J Atmos Sci 22, 3:225–261

Berkner LV, Marshall LC (1965b) Oxygen and evolution. New Sci 28, 469:415–419

Berkner LV, Marshall LC (1966) Limitation of oxygen concentration in a primitive planetary atmosphere. J Atmos Sci 23, 2:133–143

Berner RA, Lasaga AC, Garrels RM (1983) The carbonate-silicate geochemical cycle and its effect on atmospheric carbon dioxide over the past 110 million years. Am J Sci 283, 7: 641–683

Bolin B, Keeling CD (1963) Large-scale atmospheric mixing as deduced from the seasonal and meridional variations of carbon dioxide. J Geophys Res 68, 13:3899–3920

Brinkman RA (1969) Dissociation of water vapour and evolution of oxygen in the terrestrial atmosphere. J Geoph Res 74, 23:5355–5368

Bryan K et al. (1982) Transient climate response to increasing atmospheric carbon dioxide. Science 215:56–58

Buchardt B (1978) Oxygen isotope paleotemperatures from the Tertiary period in the North Sea areas. Nature 275:121–123

Budyko MI (1971) Climate and Life. Gidrometeoizdat, Leningrad, 472 (R) [English translation: Miller DH (ed, 1974) Academic Press, New York, 470 p]

Budyko MI (1972) Man's impact on climate. Gidrometeoizdat, Leningrad, 47 p (R)

Budyko MI (1974) Climatic Change. Gidrometeoizdat, Leningrad, 280 p (R) [English translation: see Budyko MI (1977c)]

Budyko MI (1977a) Global ecology. Mysl, Moscow, 328 p (R) [English translation: Progress Publishers, Moscow (1980), 323 p]

Budyko MI (1977b) Present-day climate change. Gidrometeoizdat, Leningrad, 47 p (R)

Budyko MI (1977c) Climatic Change. American Geophysical Union, Wash DC, 261 p
Budyko MI (1980) The Earth's Climate: past and future. Gidrometeoizdat, Leningrad, 352 p
   (R) [English translation: Academic Press, New York (1982), 307 p]
Budyko MI (1981) Changes in the thermal regime of the atmosphere in the Phanerozoic.
   Meteorol Gidrol 10:5−10 (R)
Budyko MI (1982) Changes in the environment and successive faunas. Gidrometeoizdat, Lenin-
   grad, 77 p (R)
Budyko MI (1984) The evolution of the biosphere. Gidrometeoizdat, Leningrad, 488 p (R)
   (English translation: Reidel, Dordrecht 1985)
Budyko MI, Ronov AB (1979) The evolution of the atmosphere in the Phanerozoic. Geochi-
   miya 5:643−653 (R)
Budyko MI et al (1978) Impendent changes in climate. Izv Akad Nauk SSSR Ser Geogr 6:5−10
   (R)
Budyko MI, Vinnikov KY, Yefimova NA (1983) The dependence of air temperature and preci-
   pitation on carbon dioxide content of the atmosphere. Meteorol Gidrol 4:5−13 (R)
Budyko MI, Ronov AB, Yanshin AL (1985) Changes in the chemical composition of the
   atmosphere in the Phanerozoic. Izv Akad Nauk SSSR Ser Geol 1:3−13 (R)
Byutner EK (1961) On the time of fixation of stable oxygen amount in the atmospheres of
   the planets containing water vapour. Dokl Akad Nauk SSSR 138:1050−1052 (R)
Byutner EK (1983) The relationship between the partial pressure of atmospheric CO_2 and the
   oceanic carbonate system. Meteorol Gidrol 10:60−67 (R)
Byutner EK (1984) On the Earth's surface temperature in the geological past. Meteorol Gidrol
   9:56−60 (R)
Callendar GS (1938) The artificial production of carbon dioxide and its influence on tempera-
   ture. Q J R Meteorol Soc 64, 27:223−240
Carbon Dioxide and Climate: a scientific assessment (1979) Nat Acad Sci, Wash DC 22 p
Carbon Dioxide and Climate: a second assessment (1982) Nat Acad Sci, Wash DC, 72 p
Chamberlin TC (1897) A group of hypotheses bearing on climatic changes. J Geol 5:653−683
Chamberlin TC (1898) The influence of great epochs of limestone formation upon the consti-
   tution of the atmosphere. J Geol 6:609−621
Chamberlin TC (1899) An attempt to frame a working hypothesis of the cause of glacial
   periods on an atmospheric basis. J Geol 7:545−584
Changing Climate (1983) Nat Acad Sci, Wash DC, 496 p
Climatic Effects of Increased Atmospheric Carbon Dioxide (1982) Proc Soviet-American
   meeting on studying climatic effects of increased atmospheric carbon dioxide, Leningrad,
   15−20 June, 1981. Gidrometeoizdat, Leningrad, 56 p (R)
Cloud PE (1974) Atmosphere, development of. Encyclopaedia Britannica, 15th edn, pp 313−
   319
Crowley TJ (1983) The geological record of climatic change. Rev Geophys Space Phys 21:
   828−877
Dobrodeev OP, Suyetova IA (1976) The living matter of the Earth. In: The problems of
   general geography and palaeogeography. Moscow State Univ, Moscow, pp 26−58 (R)
Dymnikov VP, Galin VY, Perov VL (1980) The study of climate sensitivity to CO_2 doubling
   by means of zonally averaged general circulation model. In: Mathematical modelling of the
   dynamics of the atmosphere and the ocean, part 2. Nauka, Novosibirsk, pp 39−50 (R)
Energy and Climate (1977) Studies in Geophysics Nat Acad Sci, Wash DC, 158 p
Frakes LA (1979) Climates throughout geologic time. Elsevier, Amsterdam, 310 p
Frakes LA (1984) The Mesozoic-Cenozoic history of climate change and cause of glaciation.
   27th Int Geol Congr Abstr, vol 9, part 1, pp 207−208
Garrels RM (1975) The circulation of carbon, oxygen and sulphur throughout geologic time.
   Nauka, Moscow, 47 p (R)
Garrels RM, Mackenzie FT, Hunt C (1975) Chemical cycles and global environment. Kaufman,
   Los Altos, 206 p

Garrels RM, Lerman A, Mackenzie FT (1976) Controls of atmospheric O_2 and CO_2: past, present and future. Am Sci 64:306–315

Geochemistry of Platform and Geosynclinal Rocks and Ores. In: Migdisov AA (1983) Nauka, Moscow, 263 p (R)

Geoffroy St. Hilaire E (1833) Le degré d'influence du monde ambiant pour modifier les formes animales. Mem Acad Sci Paris 12:64–93

Gilyarov MS (1975) The general trends in the evolution of insects and higher vertebrates. Zool Zh 54(6):822–831 (R)

Goldschmidt VM (1933) Grundlagen der quantitativen Geochemie. Fortschr Miner Kristallogr Petrogr 18(2):112–156

Grigoriev AA (1936) On certain interrelations among basic elements of physico-geographical environment and their evolution. Probl Fiz Geogr 3:3–30 (R)

Hallam A (ed, 1977) Patterns of evolution as illustrated by the fossil record. Elsevier, Amsterdam, 591 p

Hameed S et al. (1980)Response of the global climate to changes in atmospheric chemical composition due to fossil fuel burning. J Geophys Res 85:7537–7545

Hansen JE et al. (1979) Proposal for research in global carbon dioxide source/sink budget and climate effects. Goddard Inst Space Stud NY, 60 p

Hart MH (1975) An explanation for the absence of extraterrestrials on Earth. Q J R Astr Soc 16:128–135

Hart MH (1978) The evolution of the atmosphere of the earth. Icarus 33:23–29

Hart MH (1979) Habitable zones about main sequence stars. Icarus 37:351–357

Holland HD (1978) The chemistry in the atmosphere and oceans. Wiley, New York

Holland HD (1984) The chemical evolution of the atmosphere and oceans. Princeton Univ Press, Princeton, 582 p

Initial Reports of Deep Sea Drilling Project (1969–1981) Wash DC, vol 1–62

Khain VY, Levin LE, Tuliani LI (1982) Some of the quantitative parameters of the global structure of the Earth. Geotektonika 6:25–37 (R)

Khain VY, Ronov AB, Balukhovsky AN (1983) The Late Mesozoic and Cenozoic lithologic formations of the continents and oceans (the Early and Late Cretaceous). Sov Geol 11: 79–101 (R)

Klige RK (1980) The oceanic level in the geological past. Nauka, Moscow, 111 pp (R)

Koblents-Mishke OI, Sorokin YI (1962) The primary production of the ocean. Oceanologiya 2(2):506–510 (R)

Kondratiev KY, Moskalenko NN (1980) The evolution of the atmosphere and greenhouse effect. Izv Akad Nauk SSSR Ser Phys Atmos Ocean 16(11):1151–1162 (R)

Krasilov VA (1977) Evolution and biostratigraphy. Nauka, Moscow, 256 p (R)

Kuhn WR, Kasting JE (1983) Effect of increased CO_2 concentration on surface temperature of the early Earth. Nature 301:53–55

Lapenis AG (1984) The relationship between the carbon dioxide partial pressure in the atmosphere and the level of critical depth of carbonate accumulation in the ocean. Meteorol Gidrol 9:66–73 (R)

Li Juan-Hui (1972) Geochemical mass balance among lithosphere, hydrosphere and atmosphere. Am J Sci 272:119–137

Lisitsyn AP (1978) The processes of oceanic sedimentation: lithology and geochemistry. Nauka, Moscow, 392 p (R)

Lisitsyn AP (1980) The general principles of sedimentary layers' structure in the ocean. In: Geological history of the ocean. Nauka, Moscow, pp 36–103 (R)

Lovelock JE (1979) Gaia. Oxford Univ Press, Oxford 157 p

Luchitsky IV (1971) The fundamentals of palaeovolcanology, vol 1. Nauka, Moscow, 480 p (R)

Mackenzie FT, Pigott JD (1981) Tectonic controls of Phanerozoic sedimentary rock cycling. J Geol Soc (Lond) 138:183–196

Manabe S, Broccoli AJ (1985) A composition of climate model sensitivity with data from the last glacial maximum. J Atm Sci 42,23:2643—2651

Manabe S, Stouffer RJ (1980) Sensitivity of a global climate model to an increase of CO_2 concentration in the atmosphere. J Geophys Res 85, C10:5529—5553

Manabe S, Wetherald RT (1967) Thermal equilibrium of the atmosphere with a given distribution of relative humidity. J Atmos Sci 24, 3:241—259

Manabe S, Wetherald RT (1975) The effect of doubling the CO_2 concentration on the climate of a general circulation model. J Atmos Sci 32, 1:3—15

Manabe S, Wetherald RT (1980) On the distribution of climate change resulting from an increase in CO_2-content of the atmosphere. J Atmos Sci 37:99—118

Margulis L, Lovelock JE (1974) Biological modulation of the Earth's atmosphere. Icarus 21: 471—489

Mayr E (1976) Evolution and the diversity of life. Selected essays. Belknap Press of Harvard Univ Press, London, 721 p

Mokhov II (1981) On the CO_2 effects on the thermal regime of the Earth's climatic system. Meteorol Gidrol 4:24—34 (R)

Neftel A et al. (1982) Ice core sample measurements give atmospheric CO_2 content during past 40,000 years. Nature 295:220—223

Newman MJ, Rood RT (1977) Implications of solar evolution for the earth's early atmosphere. Science 198:1035—1037

Owen T, Cess R, Ramanathan V (1979) Enhanced CO_2 greenhouse to compensate for reduced solar luminosity on early Earth. Nature 277:640—642

Pitrat CW (1970) Phytoplankton and the late Paleozoic wave of extinction. Palaeogeogr Palaeoclimatol Palaeoecol 8:49—66

Ramanathan V et al. (1979) Increased atmospheric CO_2: zonal and seasonal estimates of the effect on the radiation energy balance and surface temperature. J Geophys Res 84, C8: 4949—4956

Raup DM, Stanley SM (1971) Principles of Paleontology. Freeman, San Francisco, 331 p

Robinson PL (1971) A problem of faunal replacement on Permo-Triassic continents. Paleontology 14, 1:131—152

Ronov AB (1949) The history of sedimentation and epeirogenic movements on the European part of the USSR (based on volumetric method). Tr Geofiz Inst Akad Nauk SSSR 3(130): 390 p (R)

Ronov AB (1959) On the Post-Cambrian geochemical history of the atmosphere and hydrosphere. Geochimiya 5:397—409 (R)

Ronov AB (1972) The evolution of the rocks' composition and geochemical processes in the Earth's sedimentary layer. Geochimiya 2:137—147 (R)

Ronov AB (1976) Volcanism, accumulation of carbon, life. Geochimiya 8:1252—1277 (R)

Ronov AB (1980) Sedimentary layer of the Earth. Nauka, Moscow, 79 p (R)

Ronov AB (1982) The global balance of carbon in the Neogäikum. Geochimiya 7:920—932

Ronov AB, Khain VY (1954) The Devonian lithologic formations of the world. Sov Geol 41: 46—76 (R)

Ronov AB, Migdisov AA (1970) The evolution of the chemical composition of the rocks of the shields and sedimentary layer of the Russian and North-American Platforms. Geochimiya 4:403—438 (R)

Ronov AB, Yaroshevsky AA (1976) New model of the chemical structure of the Earth's crust. Geochimiya 12:1763—1795 (R)

Ronov AB, Mikhailovskaya MS, Solodkova II (1963) The evolution of the chemical and mineral composition of sand rocks. In: Chemistry of the Earth's crust, vol 1. Akad Nauk SSSR, Moscow, pp 201—252 (R)

Ronov AB, Girin YP, Kazakov GA, Ilyukhin MN (1965) Comparative geochemistry of geosynclinal and platform layers. Geochimiya 8:961—979 (R)

Ronov AB, Girin YP, Kazakov GA, Ilyukhin MN (1966) Sedimentary differentiation in the platform and geosynclinal basins. Geochimiya 7:763—776 (R)

Ronov AB, Migdisov AA, Barskaya NV (1969) The principles of the development of sedimentary rocks and palaeogeographical conditions of sedimentation on the Russian Platform (an attempt at a quantitative study). Litol Poleznye Iskopayemye 6:3–36 (R)

Ronov AB, Migdisov AA, Khain VY (1972) On the reliability of quantitative methods of investigations in lithology and geochemistry. Litol Poleznye Iskopayemye 1:3–26 (R)

Ronov AB, Migdisov AA, Khain VY (1973) Possibilities and limitations of volumetric method (for the Russian Platform and the surrounding deep troughs). Litol Poleznye Iskopayemye 4:3–14 (R)

Ronov AB, Khain VY, Balukhovsky AN (1983) The Late Mesozoic and Cenozoic lithologic formations of the continents and the oceans (the Late Jurassic). Sov Geol 6:32–46 (R)

Ronov AB, Khain VY, Seslavinsky KB (1984a) The world atlas of lithologic palaeogeographical maps. The Late Pre-Cambrian and Palaeozoic on the continents. Akad Nauk SSSR, Leningrad, 70 p (R)

Ronov AB, Migdisov AA, Yaroshevsky AA (1984b) The sources of matter and the problem of the evolution of sedimentary layer and the Earth's crust. Rep 27th Int Geol Congr, Vol 11. Nauka, Moscow, pp 139–148 (R)

Rubey WW (1951) Geological history of sea water. An attempt to state the problem. Bull Geol Soc Am 62:1111–1148

Rutten MG (1971) The origin of life by natural causes. Elsevier, Amsterdam, 420 p

Sagan C (1977) Reducing greenhouses and the temperature history of Earth and Mars. Nature 269, 5625:224–226

Sagan C, Mullen G (1972) Earth and Mars: evolution of atmospheres and surface temperature. Science 177:52–56

Schmalgauzen II (1940) The ways and laws of the evolutionary process. Acad Sci USSR, Moscow, 231 p (R)

Schwab EL (1973) Geosynclinal compositions and new global tectonics. J Sediment Petrol 41, 4:928–935

Severtsev AN (1925) The principal trends of evolutionary process, 3rd edn. Moscow Univ Press, Moscow 1967, 202 p (R)

Severtsev AN (1939) Morphological features of evolution. Akad Nauk SSSR, Moscow, 610 p (R)

Shackleton NJ, Kennett JP (1975) Paleotemperature history of the Cenozoic and the initiation of Antarctic glaciation. Initial Reports of DSDP, vol 29, US Government Printing Office, Wash DC, pp 743–755

Shklovsky IS (1976) On the possible uniqueness of intelligent life in the Universe. Voprosy Filosofii 9:80–93 (R)

Simpson GG (1961) Life of the past. An introduction to paleontology. Yale Univ Press, New Haven, 198 p

Sinitsyn VM (1965) Ancient climates of Eurasia, part 1. Leningrad Univ Press, Leningrad, 267 p (R)

Sinitsyn VM (1966) Ancient climates of Eurasia, part 2. Leningrad Univ Press, Leningrad, 166 p (R)

Sinitsyn VM (1967) Introduction to paleoclimatology, Nedra, Leningrad, 232 p (R)

Sinitsyn VM (1970) Ancient climates of Eurasia, part 3. Leningrad Univ Press, Leningrad, 133 p (R)

Sinitsyn VM (1976) Climate of Laterites and Bauxites. Nedra, Leningrad, 152 p (R)

Sochava AV (1979) Changes in the composition of the Earth's atmosphere and the appearance of multicellular animals. In: Paleontology of the Pre-Cambrian and Early Cambrian. Nauka, Leningrad, pp 255–265 (R)

Sochava AV, Glikman LS (1973) Cyclic changes in the free oxygen content of the atmosphere and evolution. In: Papers of evolutionary seminar. Vladivostok, part 1, 68–87 (R)

Sokolov BS (1975) Organic world of the Earth on the way to Phanerozoic differentiation. Rep at Jubilee Session of Acad Sci USSR Moscow, izd VINITI 20 p (R)

Southman JR, Hay WW (1981) Global sedimetnary mass balance and sea level changes. In: The oceanic lithosphere, the sea, vol 7. Wiley, New York, pp 1617–1684

Takhtadzhyan AL (1980) The order: flowering plants, or Angiosperms. In: The life of plants, vol 5, part 1, Prosveshcheniye, Moscow, pp 7–114 (R)

Tappan H (1968) Primary production, isotopes, extinctions and the atmosphere. Palaeogeogr Palaeoclimatol Palaeoecol 4:187–210

Tappan H (1970) Phytoplankton abundance and Late Paleozoic extinctions. Palaeogeogr Palaeoclimatol Palaeoecol 8:49–66

Tatarinov LP (1972) Ecological factors of the origin of Amphibians. In: The problems of evolution, vol 2. Nauka, Novosibirsk, 144–153 (R)

Tenyakov VA, Yasamanov NA (1981) The Phanerozoic bauxite formations and the evolution of some of the atmosphere's parameters. Dokl Akad Nauk SSSR 257, 5:1205–1207 (R)

The World Water Balance and Water Resources of the Earth (1974) Gidrometeoizdat, Leningrad, 638 p (R)

Tikhonov AN, Lyubimov YA, Vlasov VK (1969) On the evolution of fusion zones in the course of the Earth's thermic history. Dokl Akad Nauk SSSR 188(2):342–344 (R)

Trotsyuk VY (1979) The geochemical causes of oil and gas formation in the Mesozoic and Cenozoic sedimentary layers of the World Ocean. Izv Akad Nauk SSSR Ser Geol 5:132 (R)

Tyndall I (1861) On the absorption and radiation of heat by gases and vapours and on the physical connection of radiation absorption and conduction. Phil Mag 22, 144:167–194, 273–285

Van Valen L (1971) The history and stability of atmospheric oxygen. Science 171:439–443

Vernadsky VI (1934) Essays on geochemistry. Nauka, Moscow, 380 p (R)

Vinnikov KY, Groisman PY (1981) The empirical analysis of CO_2 effects on present changes in the mean annual surface air temperature of the Northern Hemisphere. Meteorol Gidrol 11:30–43 (R)

Vinnikov KY, Groisman PY (1982) Empirical study of climate's sensitivity. Izv Akad Nauk SSSR, Phys Atmos Oceans 18, 11:1159–1169 (R)

Vinogradov AP (1959) The chemical evolution of the earth. Akad Nauk SSSR, Moscow, 44 p (R)

Vinogradov AP (1967) Introduction to the geochemistry of the Oceans. Nauka, Moscow, 215 p (R)

Vinogradov AP, Ronov AB (1956) The composition of sedimentary rocks of the Russian Platform and the history of its tectonic movements. Geochimiya 6:3–24 (R)

Walker JCG (1976) Some considerations on the evolution of the atmosphere based on the model of Earth's formation by nonhomogeneous accretion. In: The early history of the Earth. Wiley, New York

Walker JCG (1983) Possible limits on the composition of the Archaean ocean. Nature 302: 518–520

Walker JCG, Hays PB, Kasting JF (1981) A negative feedback mechanism for the long-term stabilization of Earth's surface temperature. J Geophys Res 86, C10:9776–9782

Wetherald RT, Manabe S (1981) Influence of seasonal variation upon the sensitivity of a model climate. J Geophys Res 86, C2:1194–1204

Yanshin AL (1973) On the so-called world transgressions. Bull MOIP Ser Geol 2:9–44 (R)

Yanshin AL, Zharkov MA, Kazansky YP (1977) The evolution of sedimentary rocks' formation in the history of the Earth and the associated principles of location of minerals. Geol Geofiz 11:90–97

Yefimova NA (1977) Radiative factors of the productivity of vegetation cover. Gidrometeoizdat, Leningrad, 216 p (R)

Zharkov MA (1974) The Palaeozoic salt-containing formations of the world. Nedra, Moscow, 392 p (R)

Zharkov MA (1978) The history of the palaeozoic salt accumulation. Nauka, Novosibirsk, 272 p (R)

Subject Index

Abiotic factors 107
Abyssal basin 35, 43
Actualism 19
Aerobic organisms 107, 108, 120
Air bubbles 21, 92
Air temperature 3, 4, 5, 8, 10, 11, 21, 22, 28, 29, 30, 84, 85, 88, 89, 92, 93, 95, 96, 121, 123
Albedo 10, 11, 28, 86, 90, 93, 95, 96
Algae 2, 17
Alpine cycle 45, 47
Ammonia 17, 96
Ampibians 110, 112
Anaerobic organisms 108
Ancient platforms 50, 51
Angiosperms 117
Animate nature 5, 107, 121
Anthropogenic factors 17, 121, 122, 123
Anticyclones 6
Aquatic organisms 13, 15, 17, 89
Archaean 41, 43, 45, 49, 50
Arenaceous rocks 50
Argillaceous rocks 50, 51
Aromorphosis 107, 108, 122
Atmosphere 1−7, 11, 13−24, 26, 27, 33, 45, 51, 52, 56, 63, 64, 65, 66, 70−73, 76, 77, 79, 83, 84, 94, 95, 96, 98, 100−102, 104, 105, 107, 109, 111−130
Autotrophic plants 12, 13, 14, 15, 19, 66, 92, 106, 107, 120, 121, 125, 127

Basaltic layer 35
Basaltoids 43
Basic volcanics 43, 48
Biosphere 1, 3, 5, 14, 20, 28, 29, 63, 95, 118, 119, 120, 121, 125, 129, 130
Biotic cycle 12, 13, 15, 74
Biotic environment 58, 122

Calcium 4
Caledonian cycle 45, 47
Cambrian 22, 23, 25, 74, 100, 102, 109, 110
Carbohydrates 2, 12
Carbon 3, 5, 12, 13, 14, 27, 31, 41, 43, 52, 54, 57, 63, 64, 68, 71, 72, 74, 75, 76, 99, 105, 117, 122
Carbon dioxide 1−5, 10−17, 19−31, 56, 64, 65, 66, 70, 76, 79, 80−85, 88, 91, 92, 93, 95, 96, 98, 103, 104, 105, 107, 116, 118, 120, 121, 122, 125, 127, 128, 129, 130
Carbon monoxide 3, 14, 72, 73, 76, 101, 103, 128
Carbonates 4, 13, 21, 27, 28, 30, 37, 41, 43, 49, 51, 55, 56, 57, 64, 65, 66, 68, 70, 76, 77, 83, 86, 94, 107
Carboniferous 8, 22, 24, 26, 45, 79, 80, 93, 102, 104, 109, 110−113, 116
Cenozoic 8, 9, 25, 28, 49, 50, 52, 55, 63, 73, 77, 82, 89, 89, 90, 91, 100, 102, 105
Climate 2−6, 8, 10, 66, 83, 84, 88, 89, 93, 95, 107, 116, 117, 120, 121, 123, 125, 128, 130
Cloudiness 4, 10, 28, 29, 86
Creationism 19
Cretaceous 25, 31, 35, 45, 81, 82, 102, 109, 114, 116, 117
Critical epochs 25
Cryosphere 4
Crystalline rocks 35
Cyclones 6

Earth's climate 7, 11, 66, 120
Earth's crust 2, 13, 19, 34, 35, 45, 47, 63, 64, 72, 73, 76, 77, 98, 101, 104, 107

Earth's mantle (depths)   1, 2, 3, 11, 13, 14,
    16, 20, 22, 24, 25, 26, 28, 41, 45, 51,
    52, 58, 77, 82, 97, 98, 105, 107, 114,
    118, 119, 127, 128
Echinoderms   115
Ecological crises   121
Ecological niches   116
Effusive volcanism   50, 58
Endogenetic factors   107
Endothermal vertebrates   111, 113
Eocene   22, 28, 82, 91, 116, 117, 125, 129
Epeirogenic movements   37, 58
Erosion   3, 13, 21, 34, 37, 46, 47, 48, 50
Eugeosynclinal rocks   43
Evaporation   4, 7, 123
Evaporites   37, 41, 49, 64, 105
Evolution   20, 22–29, 34, 48, 51, 52, 79,
    104, 110, 111, 112, 113, 115, 117, 118,
    125, 128
Evolutionary biology   107
Exothermal vertebrates   113

Ferruginous rocks (jaspilites)   50
Fishes   110, 115
Fossil remains   89, 113

General atmospheric circulation   6, 7, 8
Geocratic epochs   45
Geological past climates   139
Geological time scale   9
Geosynclines   35, 37, 45, 47, 50, 58
Geotectonic cycles   45, 47
Glaciations of the Earth   11, 19, 21, 29, 93,
    94, 109, 119, 129
Glomar Challenger   35
Gondwana   37
Granitic layer   35, 41, 43, 45
Granitoids   43, 48, 50
Greenhouse effect   3, 4, 5, 9, 10, 22, 27, 28,
    29, 95, 97, 119, 123

Helium   1
Hercynian cycle   45, 47
Heterotrophic organisms   3, 4, 13, 15, 17
Hydrogen   1, 14, 15, 26, 28
Hydromicaceous associations   43, 50
Hydrosphere   2, 3, 4, 12, 13, 17, 18, 28, 33,
    41, 43
Hypothesis of Gaia   118

Inert gases   1
Invertebrates   113

Jurassic   25, 35, 54, 68, 76, 81, 99, 110,
    111, 116

Krypton   1

Lavrasia   37
Liquid water   5, 8, 94, 119
Lithogenesis   18, 19, 23, 89
Lithological associations   56
Lithosphere   2, 13, 16, 18, 19, 20, 21, 28,
    47, 107, 127
Living organisms   2, 3, 4, 5, 12, 13, 15, 18,
    22, 30, 89, 94, 95, 96, 118, 120, 127

Magnesium   4
Marine organisms   89, 90, 110, 114, 115
Mathane   1, 14, 24, 123
Mesozoic   8, 27, 49, 50, 52, 73, 77, 89,
    100, 102, 104, 114, 120, 129
Metabolic level   111, 112, 113, 114, 115
Minerals   13, 17, 19, 43, 51, 72, 73, 101,
    102
Miocene   22, 28, 45, 68, 80, 82, 91, 116
Models   124
Monsoons   6
Montmorillonitic associations   43, 50
Multicellular organisms   22, 73, 108, 109

Neogäikum   37, 52, 63, 64, 67
Neogene   13, 15, 35, 89, 124
Neon   1
Nitrifying bacteria   2
Nitrogen   1, 2, 4, 5, 16, 25, 26, 27, 33, 51,
    98, 128
Nitrogen-fixing bacteris   2, 17
Noosphere   130

Ocean   6, 7, 15, 18, 22, 28, 30, 31, 41, 43,
    45, 47, 50, 51, 52, 56, 63, 68, 77, 89
    93, 121, 122
Oceanic sediments   35, 52, 54, 69, 70, 75,
    77, 88, 99
Oligocene   28, 82, 88, 91
Ontogenetic development   110
Ordovician   45, 74, 80, 102, 109, 110, 116
Oxygen   1, 2, 3, 4, 5, 12, 14–17, 19, 22–25,
    27, 51, 58, 63, 70–74, 77, 96, 100–104,
    106, 109, 110–114, 115, 117, 118, 120,
    127
Ozone   2, 3, 17, 22

Palaeoceanology   18
Palaeocene   82, 91

Palaeoclimates 89, 90
Palaeoclimatology 18, 28, 89
Palaeogene 35, 102, 109
Palaeogeographic studies 88
Palaeontology 18, 50, 51
Palaeozoic 49, 50, 51, 55, 68, 73, 100, 102, 120, 128
Parent rocks 43, 44
Permian 8, 25, 45, 81, 93, 102, 104, 113, 114, 116
Phanerozoic 8, 16, 19, 22, 23, 25, 26, 27, 33, 37, 46, 50, 51, 55, 56, 64, 65, 69, 71, 73, 74, 77, 79, 80, 81, 82, 83, 88, 89, 90, 92, 93, 94, 101, 102, 103, 104, 105, 109, 110, 113, 114, 120, 127, 129
Photodissociation 14, 15, 22, 23, 24, 28, 71
Photosynthesis 2, 3, 12, 13, 14, 17, 21, 22, 23, 24, 28, 51, 67, 68, 70, 71, 92, 104, 105, 107, 109, 116, 117, 118, 121, 128, 129, 130
Phylogenesis 111, 115
Plants 3, 4, 15, 77, 89, 92, 104, 107, 108, 109, 116, 117, 121, 125, 129
Plate tectonic theory 35
Pleistocene 88
Pliocene 13, 15, 28, 54, 68, 76, 80, 88, 99, 125
Polar climate 88
Precambrian 8, 15, 22, 24, 26, 27, 29, 30, 31, 45, 49, 50, 57, 64, 83, 92, 94, 95, 96, 97, 98, 105, 106, 108, 127, 128
Precipitation 4, 7, 123, 125
Primary atmosphere 24, 26
Progymnospermous forests 116
Proteins 2
Proterozoic 33, 35, 37, 41, 48, 49, 50, 51, 55, 63, 73, 83, 92, 93, 105, 108, 109, 113
Pterozaurs 116

Quaternary 9, 21, 35, 45, 54, 84, 88, 92, 95

Radiation equilibrium equation 10
Radiogenic heat 45
Reptiles 110, 112, 114, 116
Riphean 33
River runoff 7, 30

Sapropel 13
Sauropsids 113
Sea water 72
Secondary atmosphere 23, 26, 98, 127, 128

Sedimentary shell (stratisphere) 12, 13, 18, 35, 37, 41, 43, 47, 49, 50, 52, 54, 55, 57, 63, 64, 65, 68, 69, 73, 83, 92, 104
Seismic layers 35
Silurian 24, 25, 109
Solar constant 10, 11, 86, 87, 90
Solar radiation 4, 5, 6, 7, 9, 10, 28, 29, 86, 87, 88, 93, 94, 95, 96, 119, 129
Solar system 2, 26
Stratosphere 5, 6, 10
Subduction zone 35
Sun 6, 10, 11, 19, 21, 28, 30, 86, 118, 129

Taxonomic groups 110, 113, 115, 116, 117
Tectonic cycles 58, 117
Terrigenous rocks 37, 41, 43, 45, 47, 48, 50
Tertiary 27, 31, 68, 89, 91
Thalassocratic epochs 45
Therapsids 113, 114
Transgression of the Australian continent 46
Triassic 25, 45, 81, 102, 104, 105, 110, 111, 113, 114
Tropical climate 7
Troposphere 4, 5, 6

Ultraviolet radiation 2
Uniformitarianism 19, 127
Urey effect 14

Vendian 50, 108, 109, 110
Venus 11, 66, 95
Vertebrates 107, 110, 112
Volatiles 41, 45, 55
Volcanic activity 22, 24, 43, 57, 58, 64, 65, 66, 81, 93, 96, 97, 107, 117, 128
Volcanic rocks 37, 49, 50, 55, 57, 65, 66
Volumetrical investigation method 33

Water balance components 7
Water cycle 7, 29
Water plants 3, 12, 70
Water vapour 1, 2, 4, 5, 6, 7, 10, 14, 21, 22, 24, 26, 28, 29, 71
Weathering 4, 30, 37, 43, 45, 46, 48, 50, 51, 67, 107
"White Earth" climate 10, 11, 119
Winged insects 116
Würm glaciation 9